Quantum Physics for Beginners

From Quantum Confusion to Clarity: A Beginner's Guide to Understanding Core Concepts and Key Ideas in Quantum Physics

Orion Wells

Table of Contents

Introduction

If you've picked up this book, it's likely because you possess a natural curiosity about the inner workings of the world. You might find yourself fascinated by the grand scale of galaxies and the mysteries of black holes or equally intrigued by the minuscule, buzzing world of atoms and electrons that construct the fabric of reality. This intrigue often stems from a desire to understand the "why" and "how" behind the phenomena we observe.

Quantum physics, the study of the smallest particles and their interactions, is a key to unlocking these vast and microscopic worlds. While it might seem like a field reserved for physicists and mathematicians, it's actually centered on uncovering the principles that govern everything from the glow of distant stars to the workings of our electronic devices. However, the complex mathematics underpinning these theories often makes them daunting and inaccessible.

In this book, my mission is simple: to demystify quantum physics and make its profound concepts accessible and engaging, even if you don't consider yourself "good at math." Many times, the way scientific subjects are taught can alienate those not already well-versed in their language and methods. I believe that with relatable examples and clear explanations, anyone can grasp the principles of quantum physics and appreciate its elegance and power.

Here's a preview of what you'll discover through each chapter:

- **Chapter 1: The Quantum Breakthrough**—Uncover the origins of quantum physics and see how it diverges from classical physics. Get to know the key figures whose groundbreaking work has shaped the field.

- **Chapter 2: The Dual Nature of Light and Matter**—Explore foundational experiments like the double slit experiment and de Broglie's Hypothesis that reveal the wave-particle duality of nature.

- **Chapter 3: Quantum Superposition, Uncertainty, and Measurement**—Dive into the heart of quantum mechanics with the strange phenomena of superposition and the critical implications of the uncertainty principle.

- **Chapter 4: Entanglement and Nonlocality**—Discover the eerie world of entangled particles and the instantaneous connections that defy classical explanations.

- **Chapter 5: Quantum Tunneling—Beyond the Barrier**—Understand how particles move through and beyond barriers in ways that classical physics cannot explain.

- **Chapter 6: The Mathematics of Quantum Mechanics**—Tackle the essential mathematical tools of quantum physics, from linear algebra to complex numbers, and discover their roles in theories and equations that describe the quantum world.

- **Chapter 7: Quantum Computing and AI—The Next Frontier**—Examine how quantum mechanics is revolutionizing computing and artificial intelligence and consider the ethical implications of these technologies.

- **Chapter 8: Quantum Cryptography—The New Security Paradigm**—Learn about quantum key distribution (QKD) and how quantum principles are poised to reshape global security.

- **Chapter 9: Quantum Mechanics in Everyday Life**—See how quantum mechanics influences our daily lives, from the quantum underpinnings of photosynthesis to the applications of quantum sensors.

- **Chapter 10: Quantum Physics and Relativity—Bridging Theories**—Explore the connections between quantum mechanics and Einstein's theory of relativity through advanced concepts like field quantization.

- **Chapter 11: Quantum Paradoxes and Thought Experiments**—Challenge your understanding with discussions on quantum paradoxes and thought experiments that test the limits of human knowledge.

- **Chapter 12: The Future of Quantum Physics—What Lies Ahead**—Speculate on future discoveries and the theoretical advancements that may unravel mysteries from the nature of spacetime to the multiverse.

Each chapter in this book offers insights and explanations, along with thought challenges and reflections to help you personally connect with the material. You'll see how quantum physics underpins modern technology, medicine, and computing, profoundly transforming our understanding of the universe and driving innovations that shape our daily lives. So, let's get started and unravel the mysteries of quantum physics together!

Chapter 1:

The Quantum Breakthrough

What captivates you more: the vast expanse of the ocean or the boundless reaches of space? Perhaps it's the details of the microscopic world that hold your attention. Whether it's the macro or the micro, quantum physics plays a fundamental role in understanding both. This chapter introduces you to a world where traditional rules of physics are turned on their head and where the tiny building blocks of our universe behave in ways that challenge our everyday notions of reality.

We start our exploration with a nod to the pioneers of quantum theory—brilliant minds like Niels Bohr, Max Planck, and Albert Einstein, whose debates shaped modern understanding of the atomic and subatomic worlds. Through their stories, we'll see the fusion of relentless curiosity and rigorous scientific inquiry that has propelled the field to where it is today.

But quantum physics is more than just a series of experiments; it's a new language of mathematics and concepts that describe nature at its most fundamental level. As we move through this chapter, we won't explore deep equations yet. Instead, we'll focus on the key ideas and how they changed our understanding of the physical world. We'll see how quantum mechanics offers a different perspective than classical physics, where certainty and predictability rule.

As you read on, try to keep an open mind. The concepts of quantum physics might seem bizarre or even defy common sense, but they are the keys to understanding the fundamental components of everything in the universe. From the atoms that make up your body to the functioning of your smartphone, quantum mechanics plays a crucial role.

My goal is for you to not only have a list of facts and dates but also a real sense of wonder about quantum physics and an understanding of its basic principles. This is just the beginning, and each concept we introduce here will be explored in more detail in the chapters to come. Let's get ready to see the world in a whole new light.

From Superposition to Uncertainty: The Principles of Quantum Physics

Quantum physics is a branch of science that studies the smallest scales of the universe, revealing behaviors and phenomena that defy the laws of classical physics. While classical physics provides a framework for understanding the macroscopic world—everything from falling apples to planetary orbits—quantum physics explains the microscopic world, where particles like electrons, photons, and quarks operate.

The rules in this microscopic realm are strikingly different. In classical physics, objects have definite positions, speeds, and trajectories. They behave predictably: if you know the current state of a system, you can predict its future.

However, quantum physics introduces a world where these certainties dissolve. Particles can exist in multiple states at once—a concept known as **superposition**—until they are measured. They can also affect each other's states instantaneously over long distances, a phenomenon called **entanglement**.

A fundamental difference between classical and quantum physics is how energy is treated. Classical physics views energy as a continuum, much like a ramp one might walk up smoothly. Quantum physics, however, demonstrates that energy—and other physical properties like angular momentum and electrical charge—can only occur in fixed, discrete values. Imagine it like a staircase: you can stand on one step or another but not in between. This concept, known as **quantization**, is crucial for understanding how atoms absorb or emit energy, leading to phenomena such as the distinct color lines in the spectrum of a glowing gas.

Quantum physics also introduces the concept of **wave-particle duality**, which states that every particle or quantic entity can be partly described in terms of not only particles but also waves. This means electrons, photons, and other fundamental particles can exhibit properties of both particles and waves. For example, light can be described both as electromagnetic waves and as photons.

Let's not overlook the **uncertainty principle** formulated by Werner Heisenberg, which posits that the more precisely the position of a particle is determined, the less precisely its momentum can be known, and vice versa. This principle is not merely about the limits of measurement but indicates a fundamental property of nature. It challenges the classical idea that particles have exact, predictable positions and velocities, introducing a probabilistic approach to understanding quantum phenomena.

While seemingly theoretical, these ideas are described and predicted with precise mathematical formulas. Quantum mechanics is not simply a branch of physics; it's also deeply mathematical. These mathematical descriptions allow scientists to predict how quantum particles will behave under various conditions, leading to advancements in technology such as computers, lasers, and medical imaging devices. Thus, understanding the mathematics behind quantum physics is crucial; it's the language in which the universe's secrets are written.

Thought Challenge

The Rolling Ball vs. The Quantum Particle

Imagine a ball rolling down a hill; its path is predictable based on gravity and initial conditions, much like the principles of classical physics. Now, consider an electron moving from one point to another in an atom. Unlike the ball, the electron doesn't follow a clear path. Instead, it exists in a state of probability until it's observed—its exact position is indeterminate.

- Think about why we can ignore quantum effects when looking at something as large as a ball but not when dealing with an electron. What does this tell you about the scales at which classical and quantum physics apply?

Flipping a Coin vs. Quantum Superposition

Flipping a coin in the air gives us two clear outcomes: heads or tails. This is deterministic; if we know exactly how hard and with what spin we flip it, we can predict the outcome—classical physics in action. In contrast, a quantum bit in a computer can be both heads and tails at the same time—remember superposition—until it's measured.

- Reflect on how this idea of superposition challenges our everyday experiences. Can you think of any daily phenomena that seem to defy this either/or logic, even if they're not quantum in nature?

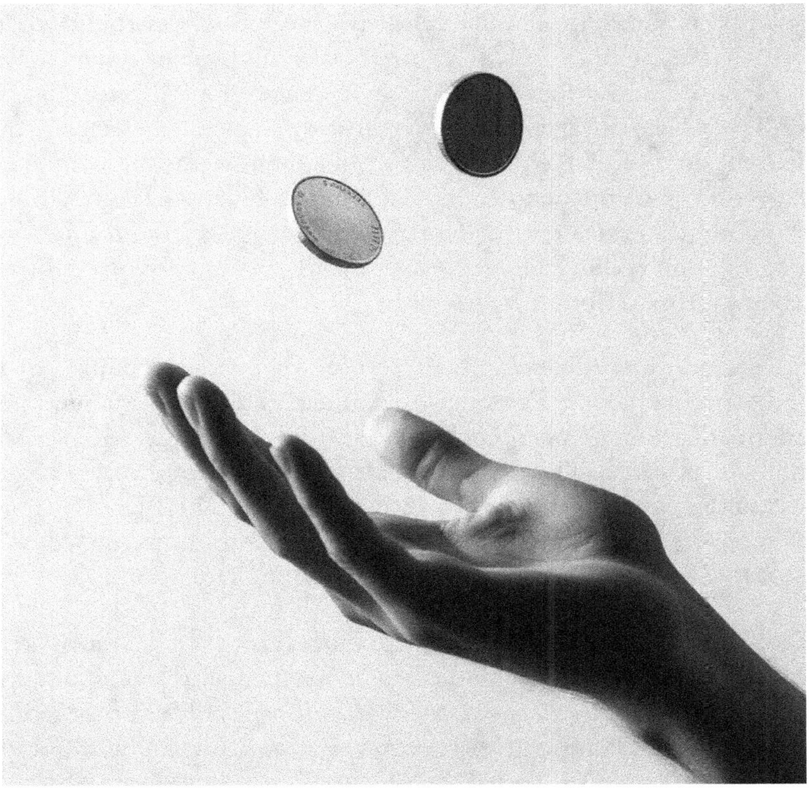

From Classical to Quantum: The Paradigm Shift

Building on these core principles, quantum physics marks a radical departure from the principles of classical physics, which had long served as the bedrock for understanding the natural world. The jcurney from classical theories to the emergence of quantum mechanics is not just a tale of scientific progress but also one of personal struggles, serendipitous discoveries, and revolutionary ideas challenging established norms.

The turn of the 20th century marked a period of profound crisis in physics. Classical physics, with its predictable and visible universe, could not explain certain phenomena. For instance, it predicted that black bodies (idealized physical bodies that absorb all incident electromagnetic radiation) in thermal equilibrium would emit radiation with infinite energy—an absurdity known as the "ultraviolet catastrophe" (Sherrill, 2006). This issue was part of a broader challenge related to black body radiation, where classical theories failed to account for the radiation's intensity across different wavelengths.

The resolution to this crisis began, somewhat unexpectedly, not with grand experiments but from the study of heat radiation—a phenomenon as mundane as the glow of coals in a campfire or the blue of a gas stove flame. As scientists like **Max Planck** explored the problem, classical assumptions were upended. On December 14, 1900, Planck proposed that energy was quantized, emitted, and absorbed in discrete packets—or **quanta**, proportional to the frequency of the radiation.

This simple yet profound adjustment, born out as an "act of desperation," as Planck himself later described, was revolutionary. It provided a quantitatively accurate formula for the color of heat radiation as a function of temperature, limiting the amount of high-frequency radiation that could be emitted at low temperatures (Styer, 1999).

This was not a moment of triumph but rather one of resignation, a last resort to salvage the consistency of existing theories. Planck's own correspondence reflects a man pushed to the edge, struggling not with elation but with the weight of a theory that would dismantle the classical framework he'd built his career upon.

Building on Planck's foundation, **Albert Einstein** advanced quantum theory in 1905 with his explanation of the photoelectric effect, demonstrating how light ejects electrons from materials and proposing that light itself is quantized into particles, which were later called photons. This not only bolstered the quantum hypothesis but also highlighted the particle-like properties of light, which were traditionally described in terms of waves—a dual nature that would become a central tenet of quantum mechanics.

The developments in quantum theory accelerated with **Niels Bohr**'s 1913 introduction of his atomic model, which employed quantization to explain atomic stability and the emission spectra of hydrogen. Bohr's model was a radical departure from classical expectations, which predicted that electrons would spiral into the nucleus and cause the atom to collapse. Instead, he proposed that electrons exist in specific orbits and transition between these orbits by absorbing or emitting quanta of energy.

Einstein and Bohr frequently interacted, and their exchanges were passionate and sometimes contentious debates that tested the limits of both friendship and professional respect. Einstein, famously skeptical of quantum mechanics' implications, once quipped to Bohr something like "God does not play dice with the universe," to which Bohr replied, "Stop telling God how to behave!" (Styer, 1999).

This ongoing debate set the stage for the Einstein-Podolsky-Rosen (EPR) paradox, a thought experiment introduced by Einstein and his colleagues in 1935 to challenge quantum mechanics on the grounds of its completeness. The EPR paradox suggested that particles could instantaneously affect each other's states across any distance, a phenomenon Einstein found unacceptable.

Bohr quickly countered, defending quantum mechanics by arguing that such non-local effects were natural consequences of the quantum world's inherent properties. This pivotal discussion highlighted key philosophical differences and foreshadowed further exploration into quantum theory's most mysterious aspects, which we will explore later in this book.

As the old quantum theory gained traction, it became clear that new mathematical tools were needed to describe the strange, non-intuitive behaviors at the atomic and subatomic levels. This led to the development of wave mechanics by **Erwin Schrödinger** in 1926 and matrix mechanics by **Werner Heisenberg** in 1925. Though initially seeming disparate, these theories were later shown to be mathematically equivalent—two sides of the same quantum coin.

Quantum mechanics did not evolve in a vacuum but was shaped by personal trials, political upheavals, and the broader social context. The First World War, personal conflicts, and even the political climate influenced the scientists involved. Theories like Schrödinger's wave mechanics and Heisenberg's matrix mechanics were forged not just in the quiet of the laboratory but also through intense personal and professional struggles, reflecting the tumultuous times.

This story, with its mix of scientific innovation and personal drama, highlights the radical shift from classical to quantum physics and the all-too-human conditions under which scientific breakthroughs are made. It reminds us that behind every great scientific advancement are the stories of individuals who dared to think differently and challenge the status quo.

Thought Challenge

Watching a Wave vs. Quantum Particles as Waves

Watch the ocean waves, and you'll see a continuous wave form moving toward the shore. Classical physics allows us to calculate the speed, wavelength, and time between waves. Now, light also behaves like a wave, but quantum mechanics shows us it's also made of particles called photons. This wave-particle duality means that light can exhibit properties of both waves and particles, depending on how we measure it.

- How does the dual nature of light challenge our classical understanding of what a wave is? Why can't we just use classical equations to describe light fully?

Heating a Rod vs. Blackbody Radiation

When you heat a metal rod, it glows red and then white as it gets hotter. Classical physics can explain this change in color through heat transfer and radiation. But at a microscopic level, this phenomenon involves emissions of photons at quantized energy levels, which classical physics can't explain on its own. Quantum physics provides a model for understanding how objects emit radiation in discrete packets.

- Consider how the introduction of quantum mechanics changes our approach to something as simple as heating an object. What does this transition from macroscopic to microscopic tell us about the role of mathematics in physics?

Reflection: The Relativity of Scale and Perspective

When we talk about what's "big" and "small," we need to remember that these terms are shaped by our human experiences—what we see, touch, and interact with. This perspective dictates when we use classical physics, suitable for the everyday world, and quantum physics, necessary for the microscopic realms that defy our usual understanding.

Imagine if we saw the world through the eyes of a different organism, for whom an atom might seem as large as a grain of sand appears to us. For such a creature, quantum effects might be ordinary, and classical physics might seem odd. This shift in perspective helps us see that "big" and "small" are not fixed; they're relative based on the observer's frame of reference.

This thought exercise invites us to step out of our human-centric view and appreciate that the universe operates on various scales, each with its own rules and mysteries. Embracing this relativity in scale enriches our understanding and reminds us of our unique place within the broader cosmic landscape.

As you ponder these ideas, consider this: How might your understanding of the world change if you viewed it from a radically different scale? What might seem mundane at one scale but miraculous at another?

Possible Answers to the Thought Challenges

The Rolling Ball vs. The Quantum Particle

Answer:

Classical physics is sufficient for everyday objects like a ball because their actions aren't significantly affected by quantum effects due to their large size and mass. Quantum effects become prominent at very small scales (like those of electrons or photons) where the particles' behavior deviates from classical expectations. This shows us that the scale at which an object operates determines whether classical or quantum physics should be applied.

Flipping a Coin vs. Quantum Superposition

Answer:

Superposition in quantum mechanics implies that things can be in multiple states at once, contrary to our daily experiences where things seem to be in one state at a time. An analogy might be thinking about potential decisions as existing in a state of superposition—until you make a decision, all possibilities exist. While not a physical manifestation, it helps illustrate the concept of multiple potential states existing simultaneously.

Watching a Wave vs. Quantum Particles as Waves

Answer:

The dual nature of light challenges our classical understanding because it shows that fundamental entities can exhibit behaviors of both waves (like interference patterns) and particles (like discrete impacts on a screen). Classical physics doesn't accommodate such duality because it traditionally treats waves and particles as separate entities. This challenges us to accept that the universe, at a fundamental level, doesn't strictly adhere to classical distinctions.

Heating a Rod vs. Blackbody Radiation

Answer:

In classical physics, heating an object is understood through the continuous absorption and emission of energy. Quantum mechanics, however, reveals that these emissions occur in quantized (discrete) packets, which explains why heated objects emit specific colors at specific temperatures. This shift to quantum mechanics introduces the need for mathematics that can handle discrete quantities and probabilities rather than the continuous changes described by classical thermodynamics.

This shift isn't just about adding complexity; it's a necessary tool to describe phenomena that classical physics can't. The mathematics of quantum physics—like probability amplitudes and wave functions—helps us describe and predict the behavior of particles at a fundamental level. How does shifting your perspective change the way you think about everything, from flipping a coin to turning on a light?

Chapter 2:

The Dual Nature of Light and Matter

Imagine stepping outside on a sunny day, feeling the warmth of the sun on your face while watching the light dance through the leaves. This simple moment encapsulates a profound scientific mystery that puzzled some of the greatest minds of the 20th century. How can light appear to wash over everything like a wave yet also deliver its energy as if it were made of particles?

In the previous chapter, we traced the evolution from classical physics, which painted a predictable, tangible universe, to the enigmatic quantum world. This chapter goes deeper, exploring the experiments and pivotal discoveries that confirmed the dual nature of not only light but all matter.

Our story picks up with a beam of light—a seemingly continuous stream that has captivated thinkers from ancient philosophers to modern physicists. Experiments, such as the famous double-slit experiment, provide startling evidence that light can exhibit both wave-like and particle-like behaviors, depending on how we observe it.

This wave-particle duality doesn't stop with light; it extends to the very building blocks of matter. Electrons, once thought to behave as tiny billiard balls circling the nucleus, also display this dual character. We'll explore how firing electrons at a barrier with two slits reveals a pattern that could only be explained if electrons, too, behaved like waves spreading out and interfering with themselves.

Each experiment we discuss is not just a step toward deeper scientific understanding but also a narrative of human curiosity and ingenuity. These discoveries challenge our everyday perceptions and invite us to think about the world in new and unexpected ways. What does it mean for something to be both a particle and a wave? How do these findings affect our understanding of the universe and our place within it?

As we explore these questions, remember the feeling of sunlight on your face and the shadows of the leaves. The beauty of nature often lies in its mysteries, and the dual nature of light and matter is one of its most intriguing.

From Atoms to Photons: Understanding the Duality

Before we explore quantum physics further, grasping the foundational concepts that underpin this field is crucial. In this exploration, we'll clarify some fundamental concepts that often cause confusion among newcomers. What do we mean by matter, light, particles, and waves? How do these relate to terms like atoms, electrons, and photons? And crucially, what exactly is dual?

Matter, in its most general sense, includes anything with mass that occupies space, encompassing all tangible objects from atmospheric air to celestial bodies. Microscopically, matter is composed of **atoms and molecules**, which are further broken down into **protons, neutrons,**

and electrons. Atoms feature a nucleus of protons and neutrons, around which electrons orbit, analogous to a solar system.

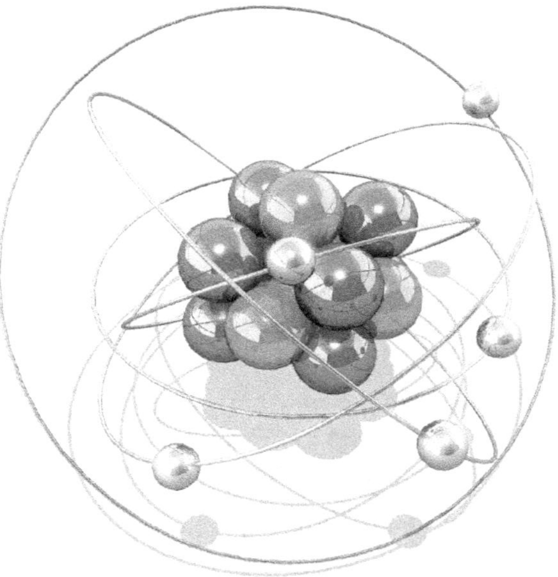

Light, traditionally viewed as a wave—a disturbance traveling through the electromagnetic field—presents characteristics like wavelength and frequency. However, it also consists of massless particles known as **photons**, embodying energy and momentum. This dual aspect of light as both wave and particle is central to quantum physics.

In classical physics, **particles** (e.g., electrons and protons) are considered distinct from waves, being discrete entities that occupy specific spatial points and follow predictable paths. **Waves**, in contrast, are expansive and capable of creating interference patterns where they overlap, intensifying or nullifying each other.

The line between particles and waves blurs in quantum physics. Electrons, though typically categorized as particles, can display wave-like behaviors such as interference, a trait previously attributed only to waves. Despite their particle-like nature, photons also show wave-like features like diffraction and interference.

Wave-particle duality is a fundamental principle in quantum physics. It indicates that all elementary particles exhibit both particle-like and wave-like characteristics. This duality, initially observed in light via the double-slit experiment, extends to all quantum particles, including matter components like electrons.

While atoms, composed of particles such as protons and electrons, generally do not show wave-like behavior due to their larger mass, under specific conditions, even they can manifest wave-like characteristics. Photons and electrons, integral to atomic structure and the behavior of matter, exemplify this duality by displaying both sets of properties.

As we explore this chapter, keep these definitions and distinctions in mind. They will help you navigate the complex but thrilling world of quantum phenomena where the only certainty is the fundamental uncertainty of everything we thought we knew about light and matter.

Thought Challenge

Sorting Through Quantum Basics

Imagine you have a set of boxes labeled Matter, Light, Particles, Waves, Electrons, and Photons. Each box is meant to hold concepts that fit the description on the label. As we explore the dual nature of light and matter, you find objects that symbolize each concept: a small rock for matter, a beam of light from a flashlight for light, tiny balls for particles, a small wave machine for waves, mini electron models, and photon emitters.

- Your task is to place each object into the correct box, considering their characteristics and behaviors in the quantum world. As you sort each object, think about which of these would also belong in a special box labeled "Dual Nature," indicating they exhibit both particle-like and wave-like properties. How does each object fit into its primary category, and which items end up in the "Dual Nature" box? This will help you visualize the distinctions and commonalities among these foundational concepts.

The Double-Slit Experiment: A Window Into Quantum Reality

Now that we've clarified the distinctions between foundational concepts, let's continue our exploration. Do you remember the groundbreaking ideas of Max Planck and Albert Einstein, which dramatically reshaped our understanding of energy and light? Planck's revelation that energy is exchanged in discrete packets, or quanta, and Einstein's demonstration that light can behave both as particles and waves set the stage for one of the most stunning illustrations of quantum mechanics: the double-slit experiment.

The Setup and Procedure

The double-slit experiment begins with a coherent light source, such as a laser, which emits light toward a barrier with two parallel slits. Beyond the barrier, a detector captures the arriving light. Classical expectations would predict two distinct bands corresponding to the slits; however, the actual result is an interference pattern of bright and dark bands, typical of wave behavior. This pattern implies that each photon passes through both slits simultaneously and even interferes with itself, contradicting any particle-only theory.

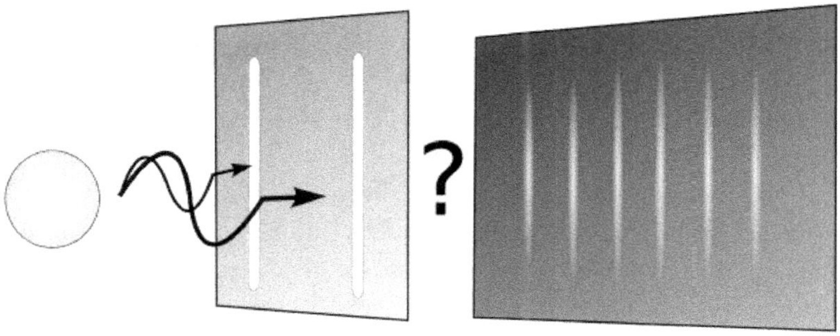

Extension to Electrons and Other Quantum Entities

The experiment becomes even more intriguing when performed with electrons. Despite being traditionally viewed as particles, electrons also produce an interference pattern, suggesting they, too, exhibit wave-like properties. This pattern forms even when electrons are fired one at a time, indicating that each electron traverses both slits simultaneously and interferes with itself, demonstrating the quintessential quantum principle of superposition.

What Does This Demonstrate About Quantum Entities?

The double-slit experiment vividly demonstrates wave-particle duality, a cornerstone of quantum mechanics. Quantum entities like photons and electrons show both particle-like and wave-like characteristics, depending on the observational setup. Unobserved, they exhibit wave-like interference; observed at the slits, their wave function collapses, revealing their particle nature. This duality underscores the principle of superposition, where a quantum system exists in multiple states until an observation forces it into one specific state.

One of the most profound revelations from the double-slit experiment is the observer effect, where the act of measuring or observing a system fundamentally alters its outcome. This phenomenon challenges classical determinism and suggests that, at a quantum level, reality is a set of probabilities that only resolve into a determined state upon observation.

The implications of this are profound, affecting both philosophical inquiries into the nature of reality and the practical applications of emerging technologies like quantum computing. The experiment forces a reevaluation of the notion that the universe behaves in a fixed, predictable manner, instead presenting a reality where possibilities hinge on the act of observation itself.

Thought Challenge

The Double-Slit Experiment and You

Imagine setting up your own version of the double-slit experiment at home. You have a small flashlight that you can use as your light source and a piece of cardboard with two parallel slits cut into it. You shine the flashlight through the slits and onto a white wall opposite. As you adjust the distance and angle of the flashlight, you start to notice patterns of light and dark on the wall.

- Based on what we've learned about the double-slit experiment, think about how you might observe similar results with your makeshift setup. What do you expect to see on the wall when you shine the light through the slits? How would your observations change if you could somehow detect which slit each photon of light passed through? Reflect on how this experiment at home could help illustrate the concepts of wave-particle duality and the observer effect in quantum mechanics.

Matter Waves and Electron Diffraction: de Broglie's Hypothesis and Further Evidence

One of the most revolutionary ideas of the early 20th century was proposed by French physicist Louis de Broglie in 1924. It fundamentally altered our understanding of the quantum world. His hypothesis—that all matter possesses wave properties—bridged the gap between the worlds of the very large and the very small and catalyzed a profound shift in how scientists perceived the nature of reality.

The Genesis of De Broglie's Hypothesis

De Broglie's hypothesis emerged from the need to explain why electrons exhibit both particle-like and wave-like characteristics. His bold assertion was that not only light (as demonstrated by photons) but all particles, including electrons, could behave like waves. This idea was inspired by the equations of Albert Einstein and Max Planck, suggesting that if light can have dual properties dependent on the context of the observation, then perhaps the same could be true for matter.

De Broglie proposed that particles such as electrons have a wavelength inversely proportional to their momentum. Here, it's essential to remember that momentum, denoted as p, is a measure of the mass and velocity of a particle, essentially describing how much motion a particle has.

This wavelength, now known as the de Broglie wavelength, is given by the equation:

$\lambda = h/p$

Where:

- λ is the wavelength.
- h is Planck's constant.
- p is the momentum of the particle.

This revolutionary idea suggested that every moving particle has an associated wave, with its wavelength inversely related to its momentum, meaning the faster a particle moves, the shorter the wavelength of its associated wave.

Experimental Verification Through Electron Diffraction

The real test of de Broglie's hypothesis came with the observation of electron diffraction, an experiment first performed by Clinton Davisson and Lester Germer in 1927. They directed a beam of electrons toward a crystalline nickel target and observed the scattered electrons on a screen.

To their astonishment, the electrons produced a pattern characteristic of wave interference, similar to the patterns produced by light in the double-slit experiment. This experiment not only confirmed the wave nature of electrons but also demonstrated the general wave-like behavior of all matter under suitable conditions.

Thought Challenge

Everyday Waves and Matter Behavior

Think about a stream of water flowing from a garden hose. When you partially cover the end of the hose with your thumb, the water sprays out in a wider, diffused pattern, much like how light diffracts when passing through a narrow slit. Now, consider this scenario in terms of de Broglie's hypothesis that matter, like electrons, can also exhibit wave-like behaviors under certain conditions.

- If you were to relate the spraying water to the behavior of electrons as described by de Broglie, how might this analogy help you understand the concept of matter waves? Consider what happens to the water's flow when the opening is narrowed (your thumb partially covering it) compared to when it is not. How does this reflect the behavior of electrons when they pass through a narrow slit and demonstrate wave-like properties?

Reflection: The Practical Implications of Quantum Dualism

As we explore the dual nature of light and matter in quantum mechanics, it's intriguing to see how these deep concepts show up in our everyday lives. Quantum physics not only improves the technology in our smartphones and makes LEDs more efficient but also reveals amazing patterns in nature. We'll discuss the technological applications more in later chapters, but it's also worth noticing how quantum mechanics helps explain the natural scenes we enjoy every day.

For example, the beautiful colors of a sunset happen because light particles, or photons, bounce off tiny molecules in the air. This scattering depends on the light's wavelength. Shorter wavelengths like blue scatter more than longer wavelengths like red, which is why sunsets often look red and orange—our eyes are better at picking up these colors.

Similarly, the brilliant colors on a peacock's feathers or a butterfly's wings aren't from pigments but from how light bends around tiny structures in the feathers and wings. This bending can either add up to make bright colors or cancel out to create darker spots, depending on how you look at them. These are everyday examples of quantum mechanics in action.

It's fascinating to think about how quantum mechanics shapes our everyday views. Next time you see a rainbow, consider what other everyday sights might be influenced by these hidden quantum rules.

What will you see differently now that you know the quantum mechanics behind it?

Possible Answers to the Thought Challenges

Sorting Through Quantum Basics

Answer:

- **Matter Box**: The small rock, a typical representative of matter, goes here as it embodies tangible objects with mass.
- **Light Box**: The beam of light from a flashlight is placed here, often described as a wave in classical physics.
- **Particles Box**: Tiny balls, symbolizing discrete entities like electrons or protons, fit well in this box.
- **Waves Box**: The small wave machine, representing phenomena that expand and overlap such as water waves, belongs here.
- **Electrons Box**: Mini electron models are placed here, categorized as subatomic particles.
- **Photons Box**: Photon emitters are assigned here, highlighting particles that are fundamental to light.

Objects in the "Dual Nature" Box:

- **Mini electron models**: While primarily particles, electrons exhibit wave-like properties like interference.
- **Photon emitters**: Photons, though particles, show wave behaviors such as diffraction.

- **Beam of light**: Initially categorized as a wave, the beam of light should also be placed in the "Dual Nature" box because it consists of photons that exhibit both wave-like and particle-like properties.

The Double-Slit Experiment and You

Answer:

In your home version of the double-slit experiment, you would expect to see an interference pattern on the wall, characterized by several light and dark bands. This pattern suggests that light acts as a wave, with the waves from each slit interfering with each other. If you could detect which slit each photon passed through (observing the photons at the slits), according to quantum theory, you would expect the interference pattern to disappear, leaving only two bright bands corresponding to the slits. This demonstrates the particle aspect of light.

Everyday Waves and Matter Behavior

Answer:

This garden hose analogy helps us visualize how matter can exhibit wave-like properties, similar to light. Just as narrowing the opening of the hose causes the water to spread out and create a spray pattern due to the change in flow dynamics, electrons passing through a narrow slit in experiments like those performed by Davisson and Germer also spread out and create patterns that indicate wave behavior.

When the hose's opening is narrowed, the water's pathway is restricted, and the flow of water fans out, creating a pattern similar to wave interference. This mimics how electrons, when confined to pass through narrow slits, display diffraction patterns indicative of waves rather than just particles.

Chapter 3:

Quantum Superposition, Uncertainty, and Measurement

Have you ever flipped a coin and, for a fleeting moment before it lands, considered that it could be both heads and tails? This everyday occurrence echoes one of the most astonishing principles in quantum mechanics: superposition. It suggests that particles, much like the coin in mid-air, can exist in multiple states simultaneously until we measure and thereby set their state.

In the last chapter, we explored the wave-particle duality of light and matter, laying the groundwork for a deeper exploration of the peculiar behaviors of quantum phenomena. Building on that understanding, this chapter discusses the strange behavior of particles, such as electrons or photons, that don't settle on a single state until we observe them. This concept is a critical function in the burgeoning field of quantum computing, which promises to revolutionize our processing capabilities by handling multiple computations at once.

We'll also unpack what happens when we measure a quantum state. The mere act of checking if the marble is in one of two boxes determines its position, collapsing its dual possibility into a single reality. This influence of measurement challenges our classical notions of object permanence and existence.

Moreover, we explore the uncertainty principle, which limits how precisely we can know certain properties about a particle, such as position and velocity. Trying to pinpoint both with perfect accuracy is akin to snapping a crystal-clear photo of a swift hummingbird in flight—a task that defies even the best technology.

These principles are not confined to theoretical physics or high-tech laboratories; they are integral to the operation of devices like MRI scanners and other electronic gadgets that define our modern existence. Let's step further into understanding how these intriguing and complex rules play out in the quantum realm and uncover just how deep the rabbit hole goes.

Exploring Quantum Superposition: From Concept to Implications

As we introduced in Chapter 1, superposition is a fundamental concept in quantum mechanics, wherein a particle exists simultaneously in all possible states until it is observed. Erwin Schrödinger initially introduced this concept in the 1920s as part of the broader development of quantum theory. Superposition isn't a theoretical abstraction but a real property observed under experimental conditions.

The classic illustration of this principle is our known double-slit experiment, where particles like electrons or photons are fired at a barrier with two slits. Instead of passing through one slit or the other, each particle passes through both slits at the same time when not observed, creating an interference pattern on a detector screen that typically would only be seen with waves. This experiment compellingly demonstrates that particles act as both particles and waves.

Imagine you have a playlist with a mix of songs from different genres. Before playing, the next song could be rock, jazz, or classical—the potential for all these genres exists in a superposed state. It's only when you hit "play" that one specific song starts, and the superposition collapses into a single definite state.

This example, much like Schrödinger's famous thought experiment involving a cat in a box that is simultaneously alive and dead until observed, illustrates the peculiar yet fundamental nature of superposition in quantum mechanics.

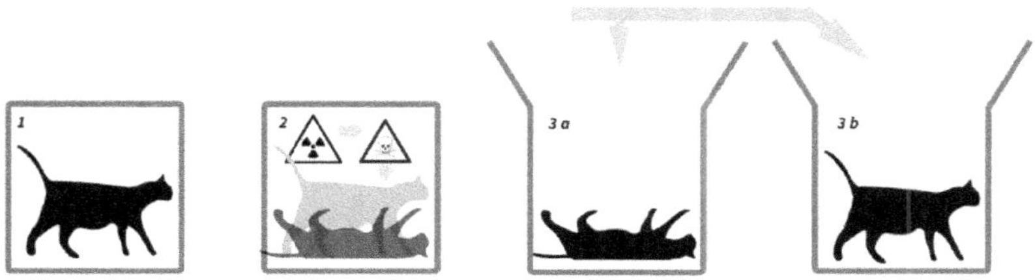

Implications of Superposition

The implications of superposition are profound and far-reaching:

- **Quantum computing:** In classical computing, a bit is a single piece of information that can either be 0 or 1. In quantum computing, quantum bits or qubits can exist in states of 0, 1, or any quantum superposition of these states. This allows quantum computers to process a vast number of possibilities simultaneously, making them potentially much more powerful than classical computers for certain tasks.

- **Encryption and cybersecurity:** Quantum superposition underpins the development of quantum encryption techniques such as quantum key distribution, which promises a level of security that is fundamentally unbreakable by conventional methods. This is because the act of measuring a quantum state (as required to intercept communication) necessarily alters that state, revealing any eavesdropping.

- **Fundamental science:** The concept of superposition allows scientists to test the limits of quantum mechanical principles and probe the boundaries between quantum mechanics and classical physics. This has implications for everything from understanding the fabric of the universe to developing new materials and technologies.

Thus, superposition remains a cornerstone of quantum physics, embodying the quantum world's strange, often counterintuitive nature. As we go deeper into these concepts, we gain a better understanding of the universe at the smallest scales and the tools to harness quantum phenomena for applications that could transform our technological landscape.

Thought Challenge: The Quantum Dice Roll

Consider a pair of dice in a box that you can shake without looking inside. Before you open the box to see the outcome, the dice could be showing any combination of numbers. This scenario is akin to a quantum system in superposition, where all possible outcomes exist simultaneously until a measurement is made.

- Relate the concept of superposition in quantum mechanics to the dice in the box. What happens when you open the box to check the outcome? How does this action reflect the quantum measurement process, where observing a particle forces it into one of its possible states?

The Uncertainty Principle: Quantum Mechanics' Limit on Precision

Following the exploration of superposition and the dual nature of quantum particles, we encounter another cornerstone of quantum mechanics: the uncertainty principle. Introduced by Werner Heisenberg in 1927, this principle states that certain pairs of properties, such as position and momentum (the product of mass and velocity), cannot both be precisely measured at the same time (Hilgevoord & Uffink, 2016).

The more accurately we determine a particle's position, the less accurately we can know its momentum, and vice versa. This isn't due to flaws in our measurement tools or techniques but is a fundamental property of the universe at the quantum level.

Heisenberg's development of the uncertainty principle was closely connected to his work on matrix mechanics, an early method of describing quantum mechanics. This approach marked a major shift from classical physics, using grids of numbers (matrices) to represent physical properties.

While working on this, Heisenberg found that when he switched the order of these matrices—which stood for the position and momentum of particles—they didn't give the same result. This mismatch showed mathematically what the uncertainty principle is about: it's impossible to know certain things about a particle, like its exact position and momentum, at the same time.

Imagine trying to locate a friend in a completely dark room using only a flashlight. The narrower the beam of light you use, the more precisely you can determine your friend's location when you spot them. However, the act of narrowing the beam increases the energy of the photons (light particles), which can push your friend slightly when they hit, thus unpredictably changing their momentum.

If you use a broader beam, you reduce the impact on your friend's momentum but at the cost of not knowing their exact location. This analogy, while simplistic, captures the essence of the uncertainty principle—observing changes observed in the quantum world.

Another way to visualize the uncertainty principle is through Heisenberg's microscope thought experiment. Imagine trying to observe an electron with a high-powered microscope. To see the electron, one must shine a light on it. However, the photons of light used to illuminate the electron will also impact it, changing its momentum. The higher the accuracy desired for the electron's position (requiring light of shorter wavelength and thus higher energy), the greater the disturbance to its momentum, illustrating the trade-off at the heart of the uncertainty principle.

Implications of the Uncertainty Principle

- **Quantum technology:** This principle plays a critical role in the design and function of quantum technologies, from the precision of quantum sensors and clocks to the reliability of quantum computers, where it informs the fundamental limits of these technologies.

- **Scientific methodology**: It has also influenced scientific methodology, highlighting the limits of observation and measurement and reminding us that our interventions as observers can alter the outcome of experiments.

- **Quantum computing**: In quantum computing, this principle influences error rates and the stability of qubits, the basic units of quantum information. Understanding and managing uncertainty is key to advancing quantum computing technologies.

Thought Challenge: The Uncertainty Principle in Action

Imagine you're at a carnival, trying to win a prize by hitting a target with a ball. The game is dimly lit, and you're wearing a pair of special glasses that blur your vision slightly, making it difficult to see the target clearly. You know that removing the glasses will give you a clear view, but doing so will also make the target move unpredictably each time you try to focus.

- Think about how this game reflects Heisenberg's Uncertainty Principle. What does the difficulty in seeing the target while wearing the glasses represent, and what does the moving target represent when you take the glasses off?

Connecting the Dots: Measuring Quantum States

The principles of superposition and the uncertainty principle are deeply interconnected and crucial for understanding how measurement affects quantum systems. These ideas converge in the act of measuring quantum states, which dramatically influences the behavior of particles at the quantum level.

Superposition allows a quantum system to exist in multiple states simultaneously. However, the moment we attempt to measure the electron's position, this superposition collapses into one of the possible states. This process directly illustrates the uncertainty principle, which stipulates that certain pairs of properties, such as position and momentum, cannot be simultaneously known to arbitrary precision.

When we measure one property (like position), we affect the other (like momentum), thus limiting our ability to predict its exact state.

Quantum Measurement and Wavefunction Collapse

Quantum measurement involves observing a particular property of a quantum system, such as the position of a particle. When this measurement occurs, the wavefunction—the mathematical expression describing the probability of a particle's quantum states—collapses. Before measurement, the wavefunction describes multiple potentialities for the particle; after measurement, it describes just one reality.

Consider an electron orbiting a nucleus. According to quantum mechanics, we can't say it occupies a specific path or location until we measure it. Before measurement, its wavefunction spreads out, describing all possible positions it might occupy. Upon measuring its position, the wavefunction collapses, and the electron is found at a specific point. This collapse happens instantaneously and unpredictably, highlighting the intrinsic randomness of quantum mechanics.

Implications of Wavefunction Collapse

The concept of wavefunction collapse has profound implications for how we understand quantum theory and reality:

- **Quantum randomness**: The collapse introduces an element of fundamental randomness in quantum theory. Unlike classical physics, where events can be predicted with certainty if all initial conditions are known, quantum mechanics only allows for probabilistic predictions.

- **Observer effect**: This randomness emphasizes the role of measurement and the observer in shaping the physical reality of quantum systems. It challenges the classical notion that objects have independent and objective properties.

- **Quantum decoherence**: The collapse of the wavefunction is key to the theory of quantum decoherence, which describes how quantum systems lose their quantum behavior and transition into classical states when interacting with the environment. This is vital for understanding why we do not observe superposition and other quantum phenomena in the macroscopic world.

We can now see a more complete picture of the quantum landscape—a world where observation not only reveals but also constructs the reality we see. These interactions between what is possible and what is observed form the backbone of modern quantum physics and continue to drive both theoretical explorations and practical applications in technology and computation.

Reflection: Quantum Mysteries and Their Influence on Reality

This chapter's journey through the intricacies of quantum superposition, measurement, and the uncertainty principle disrupts the deterministic view of the universe—a view in which the past completely determines the future if all initial conditions are known.

In the quantum world, even if we know everything possible about a system's current state, we can only predict probabilities, not certainties. This introduces an element of randomness at the most fundamental levels of physical law, suggesting that unpredictability is built into the fabric of reality.

For free will, this randomness might imply that not all events are predestined by the past but have multiple potential outcomes influenced by inherent probabilities. This perspective opens up philosophical debates about the nature of causality and choice, suggesting that, at a fundamental level, the universe may accommodate the existence of free will, as it does not strictly adhere to deterministic laws.

Considering the implications of quantum uncertainty and the role of the observer in shaping reality, to what extent do you think our lives are predetermined, and how much power do we truly have over our destinies?

Possible Answers to the Thought Challenges

The Quantum Dice Roll

Answer:

When you open the box to look at the dice, the multiple possibilities (each combination of numbers) collapse into one definite outcome—the numbers you see on the dice. This is similar to what happens in quantum mechanics during the measurement of a particle's state. Before observation, the particle exists in a superposition, embodying multiple potential states. Upon measurement, these possibilities collapse into a single state.

The Uncertainty Principle in Action

Answer:

In this carnival game scenario, wearing the glasses and having a blurred view of the target represents the uncertainty in measuring one property of a particle, like its position. When you wear the glasses, you know generally where the target is but not precisely. This is similar to knowing a particle's momentum (how fast it's moving and in what direction) but not being able to pinpoint its exact position.

Removing the glasses to see the target clearly but causing it to move unpredictably represents the uncertainty in measuring another complementary property, like momentum. When you see the target clearly (measuring the particle's position accurately), you influence the target's behavior—similar to how measuring a particle's position with high precision can disturb its momentum, making it uncertain.

Chapter 4:

Entanglement and Nonlocality

Picture this: You and a friend each hold one of a pair of "magic" dice. You travel to opposite sides of the world and agree that when one of you rolls your die, the other's die will instantly show the same number. Sounds like something from a fantasy novel, right? Yet, this is close to the reality of quantum entanglement, one of the most fascinating and mysterious phenomena in physics.

Quantum entanglement is a phenomenon where particles become so deeply linked that the state of one (no matter how far away) instantly influences the state of the other. Here, the entangled particles remain in a superposed state encompassing multiple possibilities until one is observed, at which point both particles' states are determined instantly.

In this chapter, we will explore how this intriguing aspect of quantum mechanics was first theorized and proven experimentally and how it might just redefine the future of communication and computing. We'll explore stories of how the pioneers of quantum physics grappled with the bizarre implications of entanglement, often questioning their own beliefs about the fabric of reality.

Entanglement also brings us to the doorstep of nonlocality, where things don't seem to be influenced by physical distance. This concept shakes the very foundation of how we perceive space and time and has profound implications for theoretical physics and practical applications like quantum computing and cryptography.

As we explore these complex topics, think of how interconnected and interdependent everything in the universe might be, perhaps even more than we can see or understand. What might this mean for how we connect with others, the technology we use, and our view of the cosmos?

Quantum Entanglement: The Enigmatic Bond

Quantum entanglement presents one of the most profound challenges to our classical understanding of how the universe operates. As we've seen, this phenomenon occurs when pairs or groups of particles interact in ways that make it impossible to describe the quantum state of each particle independently from the others, regardless of the distance between them. This interdependence persists even if the entangled particles are separated by vast distances—a feature Albert Einstein famously critiqued as "spooky action at a distance."

At its core, entanglement means that the properties of one particle are connected with another, allowing us to infer the state of one particle by observing its partner. For instance, if two photons are entangled in a way that their polarization states are linked, measuring the polarization of one photon instantly determines the polarization of the other, no matter how far apart they are.

The concept of entanglement challenges the classical view of the world where objects exist independently and interact through known forces over distances. In the quantum realm, entangled particles act as a single system. If you change the state of one particle, the state of the other changes in an instant, in a way that is correlated predictably despite there being no apparent communication or transfer of information between them.

The EPR Paradox and the Debate Over Quantum Completeness

The EPR paradox, named after its formulators, Einstein, Podolsky, and Rosen, in 1935, was a thought experiment intended to demonstrate what they considered a flaw in the quantum theory. The paradox outlines a scenario involving two particles that have interacted and then separated.

Remember that according to quantum mechanics, until one particle's properties are measured, they are not determined; they exist in a superposition of all possible states. When one particle is measured, the other particle's corresponding property is instantly defined, no matter the distance.

Einstein and his colleagues argued that this outcome implied that the information about the second particle's state was either communicated instantaneously, faster than the speed of light, or that the particles held hidden variables that predetermined their states from the start. This challenged the very foundations of quantum mechanics because it suggested that the theory was incomplete, potentially failing to account for these hidden variables.

Experimental Realization and Practical Production of Entanglement

In practical scenarios, entanglement can happen in different ways, usually involving light particles (photons) or electrons. For photons, one way to create entanglement involves a photon moving through a special kind of crystal that splits it into two new photons that are entangled. This means that the two new photons are connected in such a way that the state of one instantly affects the state of the other.

For electrons, entanglement can occur when electrons interact with each other in certain materials like superconductors or tiny particles called quantum dots. In these interactions, properties of electrons such as their spin (a type of intrinsic angular momentum) or their position become linked so that knowing the state of one electron gives you information about the state of the other.

Implications of Quantum Entanglement

Quantum entanglement has a range of fascinating and potentially groundbreaking implications across various fields:

- **Quantum cryptography**: Entanglement is central to quantum key distribution, a method for secure communication that allows two parties to generate a shared random secret key known only to them. The security is enhanced because any attempt to intercept the key changes the entangled state, thereby alerting the parties to the eavesdropping.

- **Quantum teleportation**: Quantum teleportation utilizes entanglement to transmit the state of a quantum system from one location to another without physically transporting the system itself, potentially leading to a secure, instantaneous transfer of information.

- **Fundamental tests of quantum mechanics**: Entanglement provides a platform for testing the limits of quantum mechanical theories, particularly through experiments that close "loopholes" in our understanding of quantum mechanics versus classical predictions.

- **Precision measurements**: Entangled particles can be used to make measurements at precision levels that are not possible with non-entangled particles. This is particularly useful in fields like spectroscopy and metrology.

- **Quantum networks**: Entanglement enables the development of quantum networks that could transform the infrastructure of data communication and connectivity, leading to a new era of information systems based on quantum technologies.

- **Understanding black holes**: Researchers use quantum entanglement concepts to explore black holes, including the famous "firewall paradox," which deals with conflicting theories in physics (Dai, 2020). The paradox suggests that instead of passing smoothly into a black hole, something falling into it would hit a wall of high-energy particles—a "firewall."

As technology progresses and more experiments are conducted, the mysteries of quantum entanglement continue to unfold, promising new scientific discoveries and a broader understanding of the universe itself.

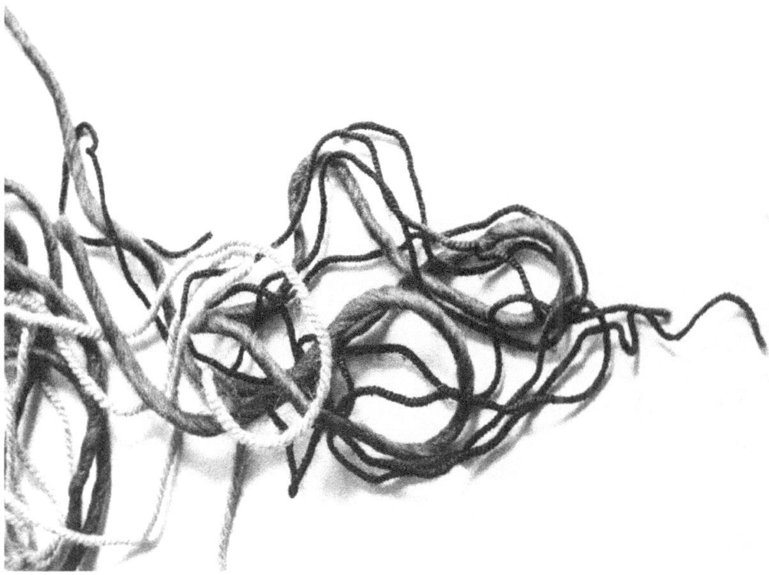

Thought Challenge: Quantum Entanglement in Daily Life

Consider that your smartphone and a friend's smartphone are capable of displaying the same notification at the exact same moment, no matter where each of you is in the world. Imagine these notifications are like entangled particles, where the action on one instantly influences the other. This scenario doesn't involve group chats but direct, personal device-to-device communication.

- How might this scenario help you grasp the concept of quantum entanglement? Reflect on how instant communication between two devices, regardless of distance, parallels the instantaneous relationship between entangled particles. What implications does this have for our understanding of communication and information transfer?

Bell's Theorem and Nonlocality: Testing Quantum Reality

Following the exploration of quantum entanglement, Bell's theorem is a critical next step in understanding the fundamental aspects of quantum mechanics. Proposed by physicist John Stewart Bell in 1964, this theorem provides a way to test whether the peculiar predictions of quantum mechanics, particularly concerning entanglement and nonlocality, are indeed real.

What Is Bell's Theorem?

Bell's theorem tackles the problem raised by the EPR paradox, which questioned whether quantum mechanics was truly a complete theory. Remember that Albert Einstein, along with his colleagues Podolsky and Rosen, famously critiqued quantum mechanics with this paradox, expressing discomfort with the idea of entanglement.

Einstein called it **"spooky action at a distance,"** arguing that it seemed impossible for two particles to influence each other instantaneously over vast distances without some hidden information being exchanged. He believed there had to be hidden variables at play, meaning that quantum mechanics was incomplete and that some deeper, local mechanism must explain these connections.

John Bell proposed a way to test this critique by formulating mathematical rules called **Bell's inequalities**. If local realism—the idea that objects are only affected by their immediate surroundings—were true, these inequalities would always hold. However, quantum mechanics predicts that under certain conditions, entangled particles would violate these inequalities. This violation would mean that changes made to one particle can instantly affect the other, no matter how far apart they are, demonstrating **nonlocality**.

Nonlocality suggests that information is somehow shared between entangled particles faster than the speed of light, directly challenging our classical understanding of causality and locality. In classical physics, effects must have causes, and these causes can only propagate at or below the speed of light. However, entanglement and Bell's theorem show that quantum particles don't follow these classical rules. Instead, they operate in a realm where distant objects can have immediate, interconnected effects on each other.

As you think about Bell's theorem, remember Einstein's famous "spooky action at a distance" and how it challenges our everyday understanding of how the universe works, where events should only affect nearby things, and information can't travel faster than light. Quantum mechanics, through entanglement and nonlocality, reveals a reality far stranger than classical physics ever imagined.

Key Experiments Testing Bell's Inequalities

Since Bell's theorem, numerous experiments have sought to test these inequalities. Alain Aspect conducted the first significant experiment in 1981-1982. Aspect and his team used entangled photons and observed that the measurement results violated Bell's inequalities, supporting the predictions of quantum mechanics and the existence of nonlocality.

These experiments have progressively closed various "loopholes" that might have allowed for local realistic explanations. For instance, the "locality loophole" suggests that entangled particles might somehow communicate their states through unknown local mechanisms if not adequately separated. The "freedom-of-choice" loophole involves the freedom in choosing measurement settings, which if restricted might bias the results. Over the years, more sophisticated tests have addressed these loopholes more rigorously.

One of the most notable recent experiments was conducted in 2015 by a team led by scientists from the National Institute of Standards and Technology (NIST) (Giustina et al., 2015). They developed a loophole-free Bell test that not only confirmed the nonlocality of quantum mechanics but also did so with such precision that it effectively closed both the locality and freedom-of-choice loopholes, providing strong support for the inherently quantum nature of the universe.

Implications of Bell's Theorem

Bell's theorem and further experiments have profound implications for our understanding of quantum mechanics and the nature of reality. They are like opening a door to a world where the rules we thought were fixed no longer apply.

- **Supports nonlocality**: The experiments show that entangled particles are directly connected, regardless of distance, supporting the concept of nonlocality, where particles affect each other faster than the speed of light.

- **Deeper understanding of entanglement**: Continued validation of Bell's theorem through experiments strengthens our understanding of quantum entanglement, confirming that quantum mechanics accurately describes these phenomena.

By leveraging entanglement, these technologies could revolutionize how we process information and secure data, making classical systems seem outdated.

Thought Challenge: Instant Connections

Think about two light switches in different rooms. You flip one switch, and at the exact same moment, the light in the other room turns on or off without any wires or connection between them. It seems like they're somehow instantly linked.

- How does this idea of instant connection relate to Bell's theorem and nonlocality in quantum mechanics? Why does this seem strange compared to how things usually work in everyday life?

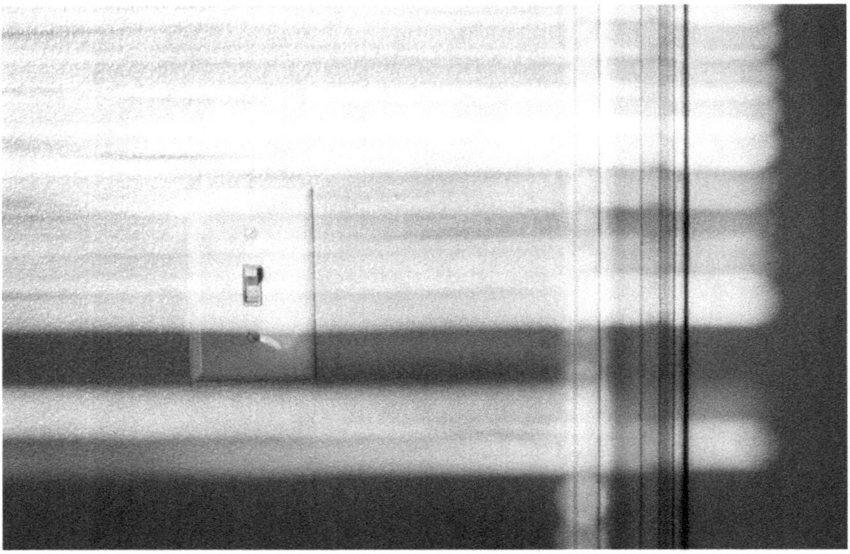

Reflection: Quantum Teleportation: Science or Fiction?

Quantum entanglement and nonlocality challenge everything we know about how the universe operates. The concept that two particles can remain instantly connected even if separated by vast distances seems to defy logic. Yet, experiments have shown that this "spooky action at a distance" is real, and it's paving the way for groundbreaking technologies like quantum teleportation.

But what is quantum teleportation? Unlike the teleportation we see in science fiction, it doesn't involve physically transporting objects from one place to another. Instead, it's about transmitting the quantum state of one particle to another instantly, regardless of distance, using entangled particles as the link.

This might sound like fiction, but it's already happening in labs around the world. Scientists have successfully teleported quantum information over distances, and while we're far from teleporting physical objects, the concept itself has exciting implications.

Think about the potential future impacts of quantum teleportation and nonlocal communication technologies. Instant communication that transcends the limitations of space could revolutionize how we interact and transmit information. Secure quantum communication channels could make traditional encryption methods obsolete, and quantum networks might one day replace our current internet infrastructure. What could this mean for privacy, international cooperation, or even space exploration?

So, is quantum teleportation just science fiction? Or is it the foundation of a future where distance is no longer a barrier? As quantum technology advances, the line between fantasy and reality may become blurrier than we ever thought possible.

Possible Answers to the Thought Challenges

Quantum Entanglement in Daily Life

Answer:

In quantum entanglement, once two particles become entangled, the state of one particle will instantly determine the state of the other, no matter the distance between them. Similarly, if smartphones were to receive notifications simultaneously, regardless of the distance between them, it would be as if the information or state of one phone instantaneously affects the other.

Instant Connections

Answer:

This scenario is similar to what happens with entangled particles in quantum mechanics. Bell's theorem shows that when we measure one entangled particle, the state of the other is instantly affected, no matter the distance between them—like flipping one switch and instantly controlling the light in another room without any physical connection. This defies our usual understanding of how things work because, in everyday life, actions typically don't have immediate effects at a distance. Bell's theorem helps explain this strange, nonlocal behavior of quantum particles, where distance doesn't limit their connection.

Chapter 5:

Quantum Tunneling—Beyond the Barrier

Have you ever thrown a ball at a wall, fully expecting it to bounce back? Imagine if, instead of rebounding, the ball simply passed through the wall as if it wasn't there. In our everyday world, this would seem impossible, but in the quantum realm, this kind of event is not only possible, it's a fundamental principle. This phenomenon is known as quantum tunneling, where particles manage to "tunnel" through energy barriers they seemingly shouldn't be able to cross.

In the classical world, barriers—whether they're physical, like a wall, or energy-based—are impenetrable if an object doesn't have enough energy to overcome them. But quantum particles don't follow these strict rules. Thanks to their wave-like nature, they have a small chance of appearing on the other side of a barrier, even without having enough energy to climb over it. It's as if the barrier was never there in the first place.

Quantum tunneling might sound strange, but it's something that happens around us all the time. It's crucial to the process that powers the sun, enabling nuclear fusion reactions to occur. Tunneling also plays a key role in modern technologies like semiconductors and flash memory, powering devices we use every day.

In this chapter, we'll explore how quantum tunneling works, where we see it in action, and how its bizarre properties have profound implications for our understanding of the universe. Though counterintuitive, this idea opens the door to countless applications, from explaining the nuclear reactions in stars to making modern electronics possible.

As we explore quantum tunneling, think about the limits you've always assumed exist. What if some of those barriers aren't as solid as they seem? Could there be other places where quantum rules apply in ways we haven't yet discovered?

The Principles of Quantum Tunneling

Quantum tunneling is one of the most intriguing and counterintuitive phenomena in quantum mechanics. It occurs when a particle passes through a potential barrier that, according to classical physics, it shouldn't be able to cross because it lacks the energy to do so. While this defies our classical intuition, quantum mechanics offers a clear explanation grounded in the wave-like nature of particles.

Wavefunction and the Quantum Mechanical Explanation

Remember that in quantum mechanics, particles such as electrons are not just point-like objects but also exhibit wave-like properties. The behavior of these particles is described by a mathematical function called the wavefunction, which provides the probability of finding a particle in a particular location.

The wavefunction doesn't abruptly stop at a barrier; instead, it extends beyond the barrier, though its amplitude decreases. This extension means there's a small probability of finding the particle on the other side, even if classically, it doesn't have enough energy to cross.

To better understand this, let's picture a ball rolling toward a hill. In classical physics, if the ball doesn't have enough energy, it rolls back. In quantum mechanics, however, the ball has a small chance of "tunneling" through the hill and appearing on the other side. This is possible because the particle's wavefunction isn't zero at the barrier—it decays but still exists, giving a non-zero probability for the particle to be found beyond the barrier.

Potential Barriers and Particle Behavior

The barrier itself is defined by its height (the energy required to cross it) and its width (how thick the barrier is). These two factors, along with the particle's energy and characteristics, determine the probability of tunneling.

- **Barrier height**: The higher the energy barrier, the less likely it is that the particle will tunnel through. This is because the particle's wavefunction decays more sharply within the barrier, reducing the probability of it appearing on the other side.

- **Barrier width**: A wider barrier also makes tunneling less likely. Even if the barrier's height remains the same, increasing the width means that the wavefunction has to decay over a longer distance, further reducing the probability of the particle being found beyond the barrier.

- **Particle energy**: The energy of the particle plays a key role in determining the tunneling probability. A particle with energy closer to the barrier's height has a higher chance of tunneling through, as its wavefunction decays less steeply compared to a particle with much less energy.

- **Particle mass**: Lighter particles, like electrons, are more likely to tunnel than heavier particles, such as protons or neutrons. The mass of the particle influences how its wavefunction behaves, with lighter particles having wavefunctions that extend further, increasing the likelihood of tunneling.

The Probability of Tunneling

The likelihood of quantum tunneling occurring depends on the interplay between the height and width of the barrier, the particle's energy, and its mass. Mathematically, the probability of tunneling decreases exponentially as the barrier's width or height increases.

Similarly, a particle with lower energy or greater mass has a reduced chance of tunneling. This explains why, in practical scenarios, tunneling is far more common for smaller particles like electrons and over very short distances, such as in atomic-scale systems.

Implications of Quantum Tunneling

Quantum tunneling is not merely a theoretical curiosity; it has observable and critical real-world applications:

- **Nuclear fusion**: Quantum tunneling is crucial in the nuclear fusion processes that power stars, including the sun. In these stellar environments, tunneling allows protons, which would normally repel each other due to similar charges, to come close enough to fuse and release energy despite not having enough energy to overcome their electrostatic repulsion.

- **Semiconductor technology**: Tunneling is fundamental to the operation of modern electronic devices. It allows for the control of electronic states in semiconductors, crucial for the functioning of transistors and diodes that form the backbone of all electronic circuits.

- **Data storage**: Quantum tunneling is key in flash memory technologies found in USB drives and SSDs. These storage devices rely on tunneling to move electrons through thin barriers, enabling the efficient writing and erasing of data.

- **Microscopy**: The scanning tunneling microscope (STM), a tool that can image surfaces at the atomic scale, operates based on quantum tunneling. It measures the tunneling current between a sharp tip and the surface of the sample, allowing for detailed atomic-level imaging.

Quantum tunneling illustrates the fundamentally probabilistic nature of quantum mechanics. The seemingly impossible act of particles moving through barriers is made possible by their wave-like properties, with the probability of tunneling determined by the barrier's characteristics and the particle's energy and mass.

Thought Challenge: Visualizing Quantum Tunneling

Imagine you're playing a video game where the main character has a "ghost mode." In this mode, the character can pass through walls and barriers as if they didn't exist, moving from one side to another without being stopped or slowed down by physical obstacles.

- How does this video game scenario help illustrate the concept of quantum tunneling? What does the character's ability to pass through barriers tell us about how particles behave in quantum mechanics?

Reflection: Quantum Tunneling—Pioneering Future Technologies

As we conclude this chapter on quantum tunneling, it's fascinating to reflect on how this seemingly esoteric phenomenon extends far beyond the confines of physics labs and into realms that might one day transform our everyday lives. Quantum tunneling isn't just about particles passing through barriers; it's a gateway to new technologies and innovations that could revolutionize industries and enhance our daily interactions with technology.

Imagine a future where quantum tunneling influences not only electronic devices but also biological processes. Scientists are already exploring how tunneling might play a role in DNA mutations or enzyme reactions—processes that are fundamental to life. The ability of particles to move through energy barriers could explain mechanisms in cellular functions that seem to defy classical physics, offering new insights into medical treatments and health diagnostics.

In the realm of consumer technology, consider how tunneling could lead to even more powerful and energy-efficient electronics. Devices could operate at speeds unimaginable today, with batteries that charge faster and last longer, all thanks to components that exploit tunneling to enhance performance and reduce power consumption.

Now, envision how these advances could reshape industry practices:

- **Medical field**: Faster, more reliable diagnostic tools that use tunneling technology could detect diseases at molecular levels long before conventional methods, offering groundbreaking shifts in early diagnosis and treatment.

- **Computing and telecommunications**: Ultra-fast, low-power computing devices and more secure communication systems could emerge, drastically altering how data is processed and protected.

- **Energy sector**: Efficient energy conversion systems that utilize quantum tunneling might significantly reduce waste and enhance sustainability, aligning with global efforts to combat climate change.

As you reflect on the vast potential of quantum tunneling, consider the broader implications of integrating quantum physics into everyday technology. What challenges might we face in harnessing and scaling up these phenomena? How might such profound scientific insights change the products you use, the medical treatments you rely on, or even the energy that powers your home?

Possible Answers to the Thought Challenge

Visualizing Quantum Tunneling

Answer:

This video game analogy helps us understand quantum tunneling by comparing the character's ghost mode to the behavior of quantum particles. In quantum tunneling, particles such as electrons can pass through potential barriers that, according to the laws of classical physics, should be insurmountable given the particle's energy level. Just like the game character in ghost mode, these particles "cheat" classical expectations by appearing on the other side of a barrier without having to climb over it.

This phenomenon occurs because quantum particles exhibit wave-like properties, and their associated wavefunctions can extend through barriers, allowing them a probability, however small, of appearing on the other side. This is akin to the video game character who can bypass walls due to their special mode, highlighting a crucial difference between our everyday experiences and the odd behaviors found at the quantum level.

Chapter 6:

The Mathematics of Quantum Mechanics

Welcome to a realm where the standard rules that govern physical objects start to unravel, revealing a universe that operates on principles and phenomena that can seem downright bizarre. Our previous discussions touched on essential mathematical tools like wavefunctions and probabilities, which are pivotal for understanding quantum phenomena. Now, we are set to explore even deeper, bringing together all the crucial mathematical concepts that serve as the foundation of quantum physics in one comprehensive chapter.

Mathematics is the language through which the universe reveals its secrets, especially in the domain of quantum mechanics. This chapter will guide you through the intricacies of linear algebra, complex numbers, and the renowned Schrödinger Equation. These concepts are the tools that physicists use to describe, manipulate, and predict the behavior of the quantum world.

If you're concerned about the complexity of these topics, rest assured. I will simplify these concepts to make them accessible and engaging. You don't need an advanced degree in mathematics to appreciate how these tools are used to unlock the mysteries of quantum mechanics.

Our goal in this chapter is to acquaint you with the mathematical underpinnings of quantum mechanics and demonstrate how these equations empower us to predict and explain phenomena that are otherwise beyond direct observation. Think of this as learning the grammar of a new language that describes everything from the tiniest particles to the vastness of space.

As we explore these mathematical concepts, remember that each formula and theorem provides deeper insights into the universe's workings. This chapter is your gateway to understanding the complex but fascinating world of quantum mathematics, where abstract symbols on a page help illuminate the most fundamental aspects of our reality.

Linear Algebra: The Quantum Framework

Linear algebra is a powerful mathematical language used in quantum mechanics to describe the states and behaviors of quantum systems. It deals with vectors, matrices, and complex spaces, making it an essential tool in understanding and predicting the behavior of particles at the quantum level.

Vectors and Matrices

In the context of quantum mechanics, vectors are used to represent the states of quantum systems. These are not just ordinary vectors you might remember from physics class, showing direction and magnitude. In quantum mechanics, state vectors are abstract mathematical objects that reside in complex vector spaces. They contain all the information about a quantum system's state, such as the position, momentum, spin, and other quantum properties of a particle.

In quantum mechanics, matrices, often referred to as operators, are used to represent physical observables like energy, momentum, or position. These operators act on state vectors, transforming them in ways that correspond to physical transformations or measurements. For example, applying a momentum operator to a state vector of a particle gives a new vector that describes the particle's momentum.

Hilbert Spaces: The Framework for Quantum States

The concept of Hilbert space is central in quantum mechanics. A Hilbert space is a complex vector space that provides a complete framework for the quantum states of a system. It's equipped with tools for calculating properties like the length of vectors and the angles between them, which, in quantum terms, relate to the probabilities of different quantum outcomes.

- **Orthogonality**: This property in Hilbert spaces refers to vectors that are perpendicular to each other. In quantum mechanics, orthogonal state vectors represent quantum states with distinct, non-overlapping properties. If you measure a quantum system described by one vector, the probability of finding it in a state described by a perpendicular (orthogonal) vector is zero.

- **Completeness**: This property means that any state vector in the Hilbert space can be expressed as a combination (superposition) of a set of basis vectors in that space. This is crucial because it allows quantum physicists to break down complex states into simpler, well-understood components.

Understanding vectors and matrices within Hilbert spaces is key to making predictions in quantum mechanics. Measurements in quantum theory are described by operators that act on state vectors within Hilbert spaces. The results of these measurements and their probabilities are determined by how these operators change the state vectors.

Moreover, the evolution of quantum systems over time is described by specific operators known as Hamiltonians, which act on the state vectors in Hilbert spaces. By applying these operators, we can predict how a quantum system will change, which is fundamental for everything from understanding the basic laws of physics to designing quantum computers.

Here's a simple table to illustrate the progression from basic mathematical concepts to the more advanced topics used in quantum mechanics:

Concept	**Basic Mathematics**	**Advanced Mathematics in Quantum Mechanics**
Numbers	Real numbers (e.g., 1, 2, 3) used for counting and basic arithmetic	Complex numbers (e.g., 3 + 2i) crucial for describing quantum states and probabilities
Geometry	Euclidean geometry dealing with shapes, lines, and angles in 2D or 3D space	Hilbert spaces, which are abstract, multi-dimensional spaces used to describe the complete state of quantum systems
Algebra	Basic algebra involving equations and inequalities with real numbers	Linear algebra involving vectors and matrices to represent quantum states and observables
Functions	Functions as mappings from real numbers to real numbers (e.g., f(x) = x^2)	Operators (special functions in quantum mechanics) that act on vectors in Hilbert spaces to change or measure quantum states
Vectors	Simple vectors in 2D or 3D space used in physics to represent forces or velocities	State vectors in complex vector spaces (Hilbert spaces) representing all possible states of a quantum system

Concept	Basic Mathematics	Advanced Mathematics in Quantum Mechanics
Matrices	Matrices for solving systems of linear equations or transformations in geometry	Operators represented by matrices that apply quantum mechanical transformations and measure quantum properties
Orthogonality	Perpendicular vectors in standard 3D space	Orthogonal state vectors in Hilbert spaces that represent mutually exclusive quantum states
Completeness	N/A in basic contexts	The ability to express any state in a Hilbert space as a sum of basis vectors, crucial for describing all possible states of a quantum system

In summary, linear algebra provides the language and tools needed to describe quantum states, make measurements, and predict the behavior of quantum systems over time. The concepts of vectors, matrices, and Hilbert spaces allow us to handle complex quantum phenomena in a precise and mathematically rigorous way. This framework deepens our understanding of the quantum world and enables the ongoing development of quantum technology.

Complex Numbers and Wavefunctions

Complex numbers are a type of number that extends the concept of the one-dimensional number line to a two-dimensional complex plane by including both a real and an imaginary component. A complex number is typically written in the form a + bi, where a is the real part, b is the imaginary part, and i is the square root of −1.

In the realm of mathematics, complex numbers are useful for performing calculations that involve square roots of negative numbers, which do not exist on the traditional number line. The algebra of complex numbers includes operations such as addition, subtraction, multiplication, and division, just like real numbers, but with the inclusion of the imaginary dimension.

Using Complex Numbers in Quantum Mechanics

In quantum mechanics, complex numbers are not just mathematical abstractions. They are essential for describing the quantum states of particles. The wavefunctions that describe these states are complex-valued functions, meaning their values are complex numbers, not just real numbers.

Wavefunctions provide a complete description of the quantum state of a system. The probability of finding a particle in a particular state or location when measured is determined by the square of the modulus (or absolute value) of the wavefunction at that point. The complex nature of wavefunctions allows them to encode more information than if they were simply real-valued, particularly about the phase relationships between different parts of the wave, which are crucial for understanding interference and diffraction patterns.

Euler's Formula and Wavefunctions

Euler's formula, which states that e^iθ = cos(θ) + isin(θ), is a pivotal expression in mathematics that links exponential functions and trigonometric functions through the imaginary unit i. This formula is central in quantum mechanics for describing the properties of wavefunctions.

In the context of quantum mechanics, Euler's formula explains how wavefunctions propagate and how their phases evolve. A typical quantum mechanical wavefunction for a particle moving in space can be expressed using Euler's formula as:

$\Psi(x, t) = Ae^{i(kx-\omega t)}$

Where:

- $\Psi(x, t)$ represents the wavefunction.
- A is the amplitude.
- k is the wave number.
- ω is the angular frequency.
- x represents position.
- t represents time.

This representation shows that the wavefunction is a complex exponential wave, where the real and imaginary parts oscillate sinusoidally (in a curve) in space and time. The parameters k and ω are related to the momentum and energy of the particle, respectively, through de Broglie's relations.

The use of complex numbers through Euler's formula allows physicists to easily calculate and predict how quantum wavefunctions evolve and interfere over time. This ability to describe the dynamic behavior of quantum systems is fundamental for all quantum mechanics, including designing experiments and developing new quantum technologies.

The Schrödinger Equation: The Quantum Master Equation

One of the most fundamental concepts in quantum mechanics is the Schrödinger Equation, which describes how the quantum state of a physical system changes over time. This equation is central to much of the work in quantum physics, providing a mathematical framework that predicts the behavior of particles at the quantum level.

Time-Dependent Schrödinger Equation

The time-dependent Schrödinger Equation addresses how the state of a quantum system evolves dynamically with time. It is expressed as (Kemp, 2022):

$$i\hbar \, \partial/\partial t \, \Psi(x, t) = H\Psi(x, t)$$

Where:

- i is the imaginary unit.
- $\hbar$ (h-bar) is the reduced Planck's constant.
- $\Psi(x, t)$ is the wavefunction of the system, depending on position x and time t.
- H is the Hamiltonian operator, representing the total energy (kinetic + potential) of the system.

This equation essentially states that the change in the wavefunction over time is dictated by the Hamiltonian, which acts on the current state of the wavefunction.

Time-Independent Schrödinger Equation

When dealing with systems where the potential energy does not change with time, we can simplify things using the time-independent Schrödinger Equation (Kemp, 2022). This form helps find the stationary states of a system, where the wavefunction's form does not change in time, only its phase does. The equation is:

$H\psi(x)=E\psi(x)$

Here:

- $\psi(x)$ is the spatial part of the wavefunction, now independent of time.
- E represents the energy eigenvalues of the system.

This equation is crucial for solving quantum systems in stable configurations, such as electrons in an atom where the potential energy is static.

Role of Mass in Quantum Mechanics

In the Schrödinger Equation, the mass of the particle is a critical component in both the time-dependent and time-independent forms, though it may not be explicitly visible in the simplified forms presented. The mass is embedded within the Hamiltonian operator (H), particularly in the kinetic energy term of the equation.

In both equations, the mass of the particle m plays a fundamental role in determining how the particle's wavefunction evolves over time and space. The kinetic energy term, which includes the mass, affects how rapidly the wavefunction's phase changes in space, a key factor in the dynamics of quantum systems. The mass, by influencing the kinetic energy term, thus impacts everything from the dispersion of the particle's wave packet to its quantum mechanical propagation through potential barriers.

Heisenberg's Matrix Mechanics: An Alternative Approach

Parallel to Schrödinger's wave mechanics, Werner Heisenberg developed matrix mechanics, which represents quantum states as matrices rather than wavefunctions. Though initially seeming quite different, it was soon shown that these two approaches are mathematically equivalent—two sides of the same coin.

In matrix mechanics, observable quantities like position and momentum are represented as matrices, and their dynamics are governed by the Heisenberg equation of motion rather than a differential equation. This form is particularly useful in dealing with quantum systems where the algebraic properties of observables (like spin) are more straightforward to manage using matrices.

Thus, the Schrödinger Equation and Heisenberg's matrix mechanics provide powerful tools for understanding and predicting the behavior of quantum systems. Whether it's the orbit of an electron in an atom or the properties of a subatomic particle like a quark, these mathematical frameworks are indispensable in the field of quantum mechanics. By translating the complex interactions of particles into a language of waves and matrices, physicists can explore and manipulate the very building blocks of the universe.

Thought Challenge: Bridging Basic Math to Advanced Quantum Mathematics

Imagine using simple arithmetic to track your daily expenses, where operations are straightforward: addition, subtraction, multiplication, and division. Now, contrast this with attempting to describe the behavior of an electron in an atom, which requires understanding its probability densities and energy states.

- How does shifting from basic arithmetic to utilizing advanced mathematical concepts in quantum physics enhance our ability to describe and predict quantum phenomena? Why do we need equations like the Schrödinger Equation instead of just using simple math?

Quantum Spin and Pauli Matrices

Both Schrödinger's equation and Heisenberg's matrix mechanics are extended to represent more complex quantum properties such as spin. Spin, a fundamental quantum property of particles, does not have a classical counterpart and can only be described within the quantum mechanical framework. Unlike properties such as mass or charge, spin is a purely quantum concept, which entails intrinsic angular momentum held by quantum particles.

Spin in quantum mechanics is represented using matrices called Pauli spin matrices. These matrices provide a way to mathematically describe the spin properties of particles like electrons, which have a spin of 1/2. The Pauli matrices are crucial for representing operations on spin states and are used extensively in quantum mechanics to handle phenomena involving spin.

The three Pauli matrices are used to express the components of spin along the x, y, and z axes, respectively. Their special mathematical properties make them suitable for describing systems where spin plays a key role, such as in the calculation of magnetic moments and in the analysis of spin-orbit interaction.

Spinors and Their Role in Quantum Mechanics

To include spin in the quantum description, physicists use spinors. A spinor is a type of mathematical object similar to a vector but designed to describe quantum states that include spin. Spinors can be thought of as an extension of wavefunctions; while a wavefunction might describe the probability amplitude of a particle's position, a spinor includes information about both the position and the spin state of the particle.

For a particle like an electron, a spinor typically has two components, each corresponding to one of the two possible spin states (spin up and spin down along a chosen axis). The overall quantum state of the particle is then a combination of its spatial wavefunction and its spinor, providing a complete description that accounts for both its position and its intrinsic spin.

Understanding the mathematical representation of spin using Pauli matrices and spinors is essential for anyone studying quantum mechanics. These tools allow physicists to describe and predict the behavior of particles with spin in various quantum systems, from atoms and molecules to solid-state devices like quantum dots and spintronic components.

Thought Challenge: Exploring Quantum Spin

Imagine you have a small, perfectly spherical magnet that can freely rotate in any direction. Now, consider if every time you tried to measure which direction the north pole of this magnet was pointing, you could only find it pointing directly up or directly down, never in any direction in between. This scenario reflects some properties of quantum spin, where particles like electrons have a spin that can only take specific, quantized orientations relative to an axis.

- How does this thought experiment with the magnet help illustrate the concept of quantum spin? What does this tell us about the quantum nature of particles compared to everyday objects?

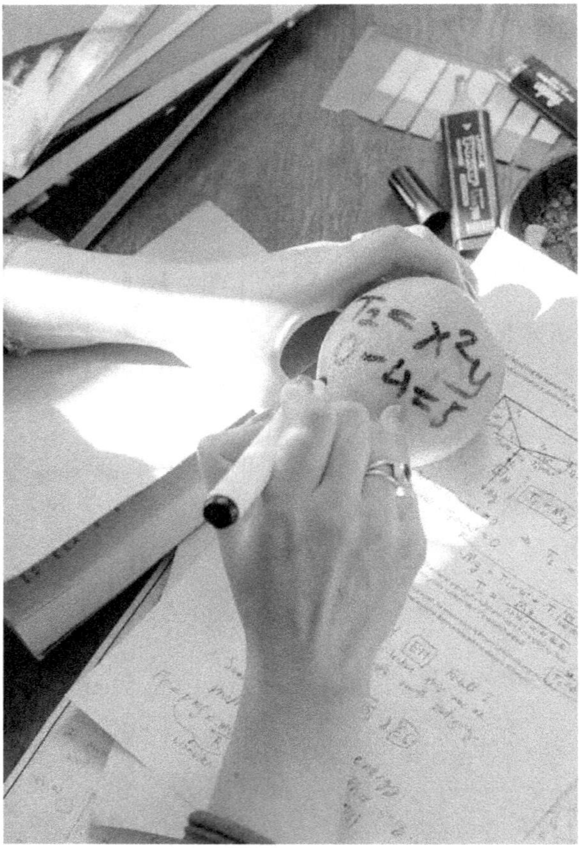

Quantum Probability and Statistics: Understanding Uncertainty in the Quantum World

A basic understanding of statistics is crucial for interpreting data and making informed decisions in every field of study. This is especially true in quantum mechanics, where probability and statistics are not just tools for analysis but fundamental aspects of the theory itself. Quantum systems exhibit inherently probabilistic behaviors, which means that understanding quantum probability and statistics is essential for anyone looking to grasp the core concepts of quantum physics.

Probability Amplitudes

In classical probability, we deal with probabilities that are direct measures of likelihood, ranging from 0 (impossible) to 1 (certain). Quantum mechanics, however, uses **probability amplitudes**, which are complex numbers associated with the probability of finding a system in a particular state. These amplitudes are not probabilities themselves; instead, the probability of an outcome is given by the square of the modulus (absolute value) of the probability amplitude.

This difference is crucial because it allows for phenomena like interference where probability amplitudes can add together, leading to probabilities that can increase or decrease depending on the phase differences between the amplitudes. This is a distinctly quantum mechanical feature that has no counterpart in classical probability theory.

The Born Rule

The Born rule is a fundamental principle of quantum mechanics that links the theory's mathematical formalism with experimental predictions. It articulates that the probability density of locating a particle in a specific state is proportional to the square of the absolute value of the wavefunction of that state. The equation is structured as follows:

$$P = |\Psi(x)|^2 dx$$

Where:

- P represents the probability of finding the particle in position x within a small interval dx.
- $\Psi(x)$ is the wavefunction at position x.
- $|\Psi(x)|^2$ is the square of the absolute value of the wavefunction, reflecting the probability density.

This rule is essential for making any physical predictions in quantum mechanics because it translates the abstract concept of wavefunctions into observable probabilities.

Statistical Description of Quantum Measurements

Quantum measurements are described statistically because of the inherent uncertainties in the state of quantum systems before measurement. When a measurement is made, the system is found in one of the possible eigenstates of the measurement operator, with the probability of each outcome determined by the wavefunction of the state prior to measurement according to the Born rule.

This probabilistic nature of quantum measurement leads to the statistical interpretation of quantum mechanics, where the outcomes of experiments can only be described in terms of probability distributions. This is different from classical measurements, where, in principle, exact outcomes can be predicted if all initial conditions are known.

Understanding quantum probability and statistics is fundamental for anyone working with quantum systems, from physicists to engineers designing new quantum technologies. The peculiarities of quantum probability, especially the role of probability amplitudes and the statistical nature of measurements, highlight the unique challenges and fascinating insights provided by quantum mechanics.

Reflection: The Mathematical Landscape of Quantum Mechanics

The mathematics of quantum mechanics provides a framework that is both powerful and challenging. It allows physicists to describe phenomena that are fundamentally beyond the scope of classical physics, from entanglement and superposition to quantum tunneling and spin. Yet, the abstract nature of these tools can sometimes seem distant from the tangible reality we interact with daily.

Here are a few questions to ponder:

- How do we bridge the gap between abstract mathematics and physical intuition? Quantum mechanics often requires us to accept principles that defy classical reasoning—how do we reconcile these with our intuitive understanding of the world?

- What challenges do physicists face when applying these mathematical models to real-world problems? For instance, how do we handle the complexities of quantum systems that involve large numbers of particles or the integration of quantum theory with theories of gravity?

- How does the abstract nature of quantum mathematics influence our understanding and interpretation of the universe? Does it limit our understanding to what is mathematically expressible, or does it perhaps open up new avenues for grasping the nature of reality that we couldn't perceive before?

These tools are not just mathematical constructs but reflections of the underlying realities of our universe. They challenge us to think differently about the nature of knowledge and observation, pushing the boundaries of what we consider possible.

Possible Answers to the Thought Challenges

Bridging Basic Math to Advanced Quantum Mathematics

Answer:

In basic math, you might use equations for simple calculations, like balancing a budget:

Total Expenses = Food + Rent + Utilities

This equation is linear and deterministic. Every term and result is directly observable and measurable. However, in quantum physics, things aren't as straightforward because particles behave in ways that defy classical intuition. For example, the Schrödinger Equation, which is fundamental in quantum mechanics, is used to determine the probability amplitudes of a particle's position and other properties.

Unlike the straightforward equation for expenses, the Schrödinger Equation uses complex numbers and derivatives to describe phenomena that are not directly observable—like the position of an electron—which can only be described probabilistically.

The use of complex numbers, partial derivatives, and operators in quantum mechanics allows for the description of systems that are influenced by factors such as superposition and entanglement. Basic arithmetic lacks the structure to accommodate these concepts because it assumes that quantities are definite and directly measurable.

Thus, the transition to using advanced mathematical tools and equations in quantum physics is essential for accurately describing and manipulating the quantum world. These tools allow us to move beyond the limits of classical physics, providing a deeper understanding of the universe's fundamental nature and supporting the development of technologies like quantum computers and secure communication systems.

Exploring Quantum Spin

Answer:

This thought experiment highlights the quantized nature of quantum spin, mirroring how subatomic particles behave in ways that are fundamentally different from macroscopic objects. In classical physics, objects like our hypothetical magnet can point in any direction with a continuous range of possibilities. However, in quantum mechanics, the spin of particles like electrons is quantized, meaning it can only exist in certain discrete states (e.g., "spin up" or "spin down" along a measured axis).

This difference is significant because it shows that quantum spin doesn't behave like any classical type of angular momentum or magnetism. Instead, it's a purely quantum phenomenon that does not have a direct classical analogy. The behavior of quantum spin, including its representation through Pauli matrices and its description by spinors, underscores the strange, non-intuitive nature of quantum mechanics.

This quantized behavior influences everything from the magnetic properties of materials to the way quantum computers might perform calculations, demonstrating the profound implications of quantum mechanical principles in both natural phenomena and technological applications.

Chapter 7:

Quantum Computing and AI—The Next Frontier

Standing on the brink of a technological revolution, we are witnessing the emergence of quantum computing, a concept that promises to reshape our understanding of computational power and problem-solving.

Throughout our exploration of quantum physics, we have found intriguing phenomena such as superposition, entanglement, and tunneling and their implications. These principles, once theoretical puzzles, are now the building blocks for innovations that could dramatically enhance our capabilities across various fields.

This chapter explores how the principles of quantum mechanics are being applied to develop groundbreaking technologies. Quantum computing, in particular, diverges significantly from traditional computing. Classical computers use bits as the basic unit of data, which are strictly binary. In contrast, quantum bits or qubits operate in states that encompass numerous possibilities simultaneously, offering potential computing power that is exponentially greater.

We'll also examine quantum gates, which manipulate qubits through fundamental quantum operations. These gates are the building blocks of quantum circuits, leading us to the development of complex algorithms capable of solving problems beyond the reach of classical computing. One such algorithm is Shor's Algorithm, renowned for its potential to factor large integers exponentially faster than the best-known classical algorithms, posing a profound challenge to current cryptographic systems.

Additionally, we will touch on the emerging field of quantum artificial intelligence (AI), where quantum computing meets machine learning. Quantum AI promises to harness the probabilistic nature of quantum mechanics to enhance machine learning models, potentially revolutionizing areas like drug discovery, climate modeling, and financial analysis.

As we traverse through these topics, we'll reflect on the broader implications of these technologies. How will quantum computing redefine our approach to problem-solving in the digital age? What challenges and opportunities might arise as we transition from classical to quantum technologies in sectors like cybersecurity, telecommunications, and beyond?

Introduction to Quantum Computing: Understanding Qubits and Quantum Gates

Quantum computing represents a seismic shift from classical computing, rooted in its fundamental unit of information: the qubit. Unlike classical bits, which encode information as either 0 or 1, a qubit can be in a state of 0, 1, or any quantum superposition of these states.

This capability allows qubits to perform complex calculations that involve large datasets much more efficiently than classical bits. For example, a quantum computer with several qubits can process a multitude of potential inputs at the same time, thanks to superposition.

Additionally, qubits are interconnected by quantum entanglement, which allows the state of one qubit to depend on the state of another, no matter the distance between them. This entanglement is key to the quantum computer's speed and power, as operations on one qubit can instantaneously influence others entangled with it, vastly accelerating the computation process.

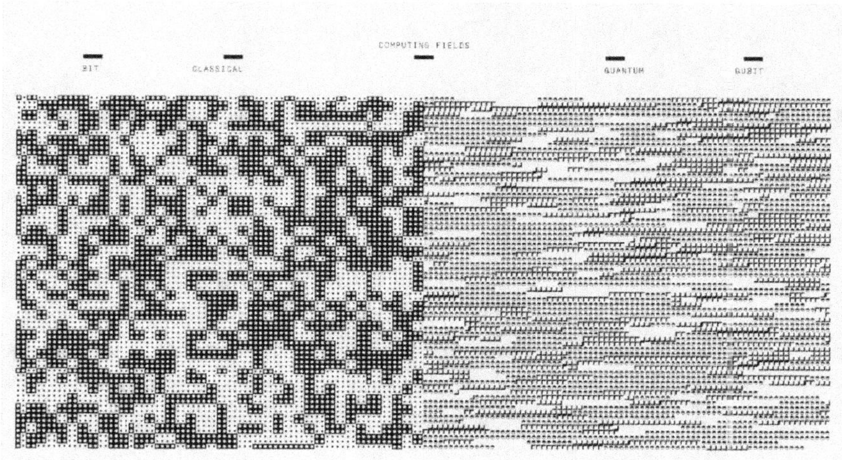

Quantum Gates and Circuits

In classical computing, bits are manipulated using logic gates such as AND, OR, and XOR to perform computations. Quantum gates manipulate qubits and are the building blocks of quantum circuits. These gates operate on a more complex set of rules due to the properties of qubits. For instance, while a classical AND gate outputs a 1 only if both its inputs are 1, a quantum gate like the Controlled-NOT (CNOT) might entangle two qubits or flip the state of one qubit depending on the state of another.

Quantum gates operate through unitary operations, which are reversible transformations. This is a fundamental difference from many classical operations, which can lose information (e.g., AND gate from two 1s gives 1, but the reverse operation isn't uniquely determined from the output alone). The reversibility in quantum gates ensures that quantum computations can be undone and retraced, a crucial feature for quantum error correction.

A quantum circuit is composed of a series of quantum gates designed to perform a specific function; it represents the entire computation as a sequence of steps. Each gate manipulates one or more qubits, performing operations that together solve problems. These circuits are typically designed to exploit quantum phenomena such as superposition and entanglement to perform tasks that would be either impossible or impractically time-consuming on classical computers.

Thought Challenge: Visualizing the Power of Quantum Computing

Imagine you have two boxes, each capable of storing a single marble that can be either red or blue. In classical computing, if you were to check each box, each would definitely hold one marble of a specific color—red or blue. Now, envision a quantum scenario where each box is like a qubit, and instead of a single color, each marble can be both red and blue at the same time until you look inside.

- How does this scenario illustrate the difference between classical computing and quantum computing? What does the ability of the marbles (qubits) to be both colors simultaneously suggest about the potential capabilities of a quantum computer?

Quantum Algorithms: Unlocking New Capabilities

Quantum algorithms are sets of instructions designed for quantum computers. These algorithms leverage the unique properties of quantum bits (qubits), such as superposition, entanglement, and quantum interference, to perform calculations that are either extremely slow or impossible for classical computers. Let's explore some of the groundbreaking quantum algorithms that illustrate the profound impact of quantum computing.

Shor's Algorithm

One of the most famous quantum algorithms is Shor's Algorithm, developed by mathematician Peter Shor in 1994. This algorithm revolutionizes the task of factoring large numbers, which is a cornerstone of modern cryptography. Classical algorithms require computational time that grows exponentially with the number of digits in the number to be factored, making it impractical to factor very large numbers efficiently.

Shor's Algorithm, however, can factor these numbers in polynomial time, drastically reducing the time needed to break down large integers into their prime factors. This capability has profound implications for security because many encryption systems, such as RSA, depend on the difficulty of factoring large numbers as the basis for their security.

If a quantum computer running Shor's Algorithm becomes practical and scalable, it could potentially decrypt many of the cryptographic systems currently in use, necessitating a complete reevaluation of our digital security strategies.

Grover's Algorithm

Developed by Lov Grover in 1996, Grover's Algorithm demonstrates how quantum computing can speed up searches through unsorted databases. In the best case, classical search algorithms need to check each element one by one, which means the time to find the target grows linearly with the size of the database. Grover's Algorithm, however, can find the target in a database of N items in roughly $\sqrt{N}$ steps, providing a quadratic speedup.

This quadratic speedup is significant for tasks involving searching and optimization problems in large datasets, such as finding a specific piece of information in a large database or solving NP-complete problems more efficiently than classical algorithms can. While not as dramatic as the exponential speedup offered by Shor's Algorithm, Grover's enhancement can still provide substantial improvements for many practical problems.

Quantum Simulation

Beyond specific algorithms like Shor's and Grover's, quantum computers hold the promise of simulating complex quantum systems directly. Many systems in physics, chemistry, and materials science are governed by quantum mechanics, and simulating these systems on classical computers often requires approximations that limit the accuracy of the results.

Quantum computers can, in principle, simulate quantum systems naturally and exactly because they themselves operate according to quantum mechanics. This capability could transform fields such as drug discovery by allowing scientists to model molecular interactions at an unprecedented level of detail or materials science, where quantum simulations could lead to the discovery of new materials with desired properties.

Quantum AI: The Frontier of Artificial Intelligence and Quantum Computing

AI has evolved dramatically over the past few decades, transitioning from basic algorithms that automate simple tasks to complex machine learning (ML) systems capable of driving cars, diagnosing diseases, and even writing human-like text. Classical computers largely power these advancements. However, as AI continues to push the boundaries of what's possible, the limitations of classical computing become more apparent, especially for tasks requiring immense computational resources or dealing with highly complex data.

The Role of Quantum Computing in AI

Quantum computing holds the potential to significantly enhance machine learning and AI by leveraging the inherent advantages of quantum mechanisms, such as superposition and entanglement. These properties can be used to process information in a fundamentally different way from classical computers, enabling more efficient algorithms.

- **Enhanced computational speed:** Quantum computers can potentially perform calculations at speeds unattainable by classical computers, significantly speeding up the training times for machine learning models, especially those involving vast amounts of data and complex calculations.

- **Handling high-dimensional data:** Quantum systems naturally operate in high-dimensional Hilbert spaces, making them potentially excellent at handling high-dimensional data, which is common in advanced AI applications.

- **Optimization problems:** Many AI and ML tasks, such as training neural networks or solving resource allocation problems, boil down to complex optimization problems. Quantum algorithms, such as the quantum approximate optimization algorithm (QAOA), are designed to find optimal solutions more efficiently than classical algorithms.

Proposed Applications of Quantum AI

Quantum AI is anticipated to revolutionize several areas by providing new solutions to old problems and enabling new types of applications:

- **Pattern recognition:** Quantum algorithms can potentially improve pattern recognition systems by enhancing the feature detection capabilities in complex datasets, such as those found in facial recognition or anomaly detection in security systems.

- **Optimization in logistics and scheduling:** Quantum-enhanced optimization could transform industries like transportation and logistics, where AI is used to optimize routes and schedules. For instance, a quantum algorithm could quickly calculate the most efficient routes for delivery trucks in a city, considering all possible variables and constraints.

- **Drug discovery:** Quantum AI could accelerate the discovery of new drugs by enabling the simulation and analysis of molecules at a quantum level, which would be prohibitively expensive and time-consuming with classical computing resources.
- **Financial modeling:** Quantum AI can be used to optimize portfolios and manage risk by analyzing historical data and simulating different economic scenarios under complex financial models.

Challenges and Future Outlook

While the potential for quantum AI is immense, significant challenges remain. Current quantum computers have limitations, such as error rates and coherence times, which must be overcome before they can be effectively used for practical AI applications. Moreover, the development of quantum algorithms that outperform classical counterparts is still in its early stages.

As researchers continue to explore these challenges, integrating quantum computing with AI promises to create more powerful, efficient, and capable AI systems. This integration, often referred to as quantum AI, represents a thrilling intersection of two cutting-edge technologies, each poised to change the world in its own right.

As we progress, the synergy between quantum computing and AI is likely to be a significant area of research and innovation, potentially leading to breakthroughs that today seem like the realm of science fiction.

Reflection: Ethical Implications of Quantum Computing and AI

The potential of quantum computing to solve problems that are currently intractable for classical computers means that sectors such as pharmaceuticals, materials science, and logistics could see revolutionary changes. For instance, quantum computing could enhance risk assessment and fraud detection, providing more robust economic systems. However, it could also disrupt current encryption methods, raising significant security concerns that must be addressed to protect data privacy and financial stability.

So, the power of quantum AI also raises substantial ethical considerations. As with any technology, there is potential for misuse, and the capabilities of quantum computing could amplify both the benefits and risks associated with AI. Issues of bias, privacy, and security become more complex in the context of quantum AI, where decisions may be made based on calculations that are difficult to audit or understand due to the probabilistic nature of quantum algorithms.

How we address these challenges, integrate ethical considerations into the development of quantum technologies, and prepare for potential disruptions will be crucial. Ensuring these technologies are developed responsibly and inclusively could help mitigate risks while maximizing benefits.

Moreover, the democratization of technology—ensuring fair access to the benefits of quantum AI—is vital. As we stand on the brink of this new frontier, it is essential to consider not only what quantum computing and AI can do but also who they do it for and who might be left behind.

As you reflect on the transformative potential of quantum computing and AI, consider the broader implications for society, the economy, and ethics. What roles will you and your community play in shaping this future? How can we ensure that the development of these technologies benefits humanity broadly, addressing global challenges and improving lives while safeguarding fundamental values?

Possible Answers to the Thought Challenge

Visualizing the Power of Quantum Computing

Answer:

This scenario serves as a metaphor to highlight the principle of superposition in quantum computing. In classical computing, bits must be in one of two definite states (0 or 1, a red or blue marble in our analogy). Each bit or box is independent, and its state is fixed and unambiguous when observed.

In quantum computing, however, qubits like our marbles can exist in multiple states at once (both red and blue simultaneously) thanks to superposition. This means that a quantum computer with just two qubits can simultaneously represent four possible states (red-red, red-blue, blue-red, blue-blue) instead of being limited to just one state at a time, as in classical computing.

The capacity of qubits to exist in multiple states exponentially increases the information density they can represent, allowing quantum computers to process a vast number of potential outcomes simultaneously. This is crucial for solving complex problems that involve a large number of possible combinations and is especially powerful in fields like cryptography, optimization, and simulation of molecular structures.

Chapter 8:

Quantum Cryptography—The New Security Paradigm

During World War II, the Enigma machine stood as a formidable symbol of encrypted communication. The codebreakers at Bletchley Park, led by the renowned Alan Turing, embarked on a relentless quest to decipher the German codes. Their success changed the course of the war and marked the beginning of modern cryptography. Fast forward to today, and we find ourselves on the brink of another monumental shift in the world of cryptography, propelled this time by the principles of quantum mechanics.

In previous chapters, we explored the mind-bending realms of quantum mechanics, from particles being in two places at once to the entanglement phenomenon where particles, no matter the distance between them, appear to communicate instantaneously. These principles are the keystones of a new era of cryptography—quantum cryptography.

As we venture into the nuances of this revolutionary approach, we'll explore how quantum key distribution (QKD) leverages the unique aspects of quantum mechanics to offer security rooted in the fundamental laws of physics, surpassing the boundaries of human creativity and computational capabilities.

Imagine a world where your personal and financial information, private communications, and even state secrets are secured by the principles governing the subatomic particles. This isn't a script from a sci-fi movie; it's a reality unfolding as quantum cryptography begins to influence global data security protocols. As we unravel the complexities and implications of securing information in the quantum era, think about the enormity of shifting from mechanical gears and paper strips of Turing's time to the quantum states of photons.

How will society adapt to a new standard of unbreakable encryption? What challenges and transformations should we anticipate as quantum security practices become mainstream?

Principles of Quantum Cryptography: Securing Information

Quantum cryptography, at its core, uses the properties of quantum mechanics to secure data. The most prominent feature of quantum cryptography is QKD, which allows two parties to generate a shared random secret key known only to them, which can then be used to encrypt and decrypt messages. The security of QKD arises from two fundamental quantum phenomena we discussed:

- **Quantum superposition**: In quantum mechanics, particles like photons can exist simultaneously in multiple states. Quantum cryptography uses this principle to encode information in quantum states in ways that are inherently secure from eavesdropping.

- **Quantum entanglement**: When particles become entangled, the state of one particle instantly correlates with the state of another, no matter the distance between them. This property is used in quantum cryptography to detect any attempt at eavesdropping, as measuring one particle's state will invariably alter the state of its entangled partner.

Another critical concept in quantum cryptography is the **no-cloning theorem**, which states that it is impossible to create an identical copy of an arbitrary unknown quantum state. This theorem underpins the security of quantum communication methods, as any attempt to intercept and measure the quantum key alters its state, thus alerting the legitimate parties to the presence of an eavesdropper.

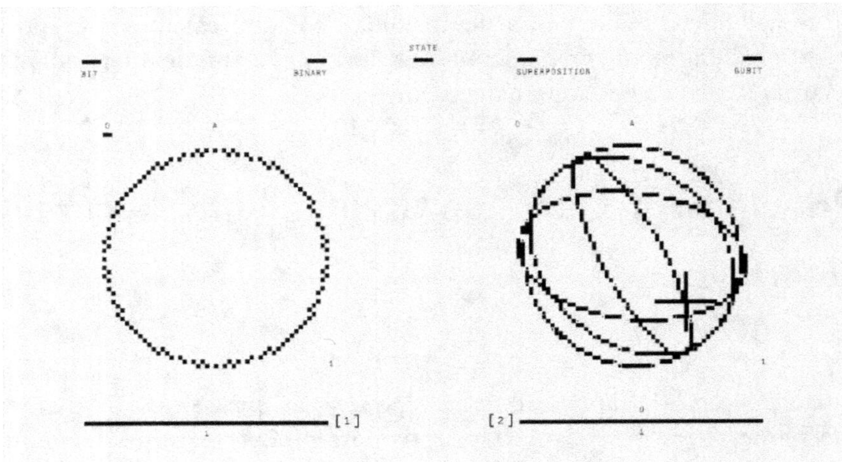

QKD Protocols

QKD protocols, such as BB84 and E91, use these quantum properties to ensure secure communication. Here's how they typically work:

- **BB84 protocol**: Developed by Charles Bennett and Gilles Brassard in 1984, this protocol involves the sender (Alice) sending photons polarized in one of four possible orientations to the receiver (Bob). The orientations are chosen such that they represent bits of information in two mutually unbiased bases. Bob randomly chooses the basis to measure the incoming photons. After the transmission, Alice and Bob publicly compare their choice of bases (without revealing the outcome of their measurements) and keep only the results from matching bases, thus creating a secure key.

- **E91 protocol**: Proposed by Artur Ekert in 1991, this protocol relies on entangled photon pairs. Both Alice and Bob receive one photon from each pair and perform measurements. The correlations predicted by quantum mechanics between their measurements confirm the security of their key against eavesdroppers.

Contrast with Classical Cryptography

Classical cryptography, including widely used methods like RSA and AES, relies on the computational difficulty of certain mathematical problems (e.g., factoring large integers and solving discrete logarithms). The security of these systems depends on the limits of current technology and mathematical techniques; sufficiently powerful classical or quantum computers could potentially break them.

In contrast, quantum cryptography is not based on the difficulty of mathematical problems but on the fundamental laws of quantum physics. This difference means that the security of quantum cryptography is theoretically unbreakable as long as the laws of quantum mechanics hold, offering a level of security that classical systems cannot match. This aspect makes quantum cryptography particularly appealing in an era where quantum computing poses a significant threat to classical cryptographic systems.

By integrating the peculiarities of quantum behavior, quantum cryptography introduces a new paradigm in secure communications, ensuring that the future of data protection is deeply rooted in the unbreakable laws of physics.

Current Technologies and Developments

Quantum cryptography is progressing from theoretical exploration to practical implementation. Let's examine the current technologies fueling these advancements, examine ongoing global projects, and discuss the challenges faced in the field.

Hardware Used in QKD

- **Single-photon detectors**: These devices are crucial for QKD as they detect single photons that carry the quantum information necessary for generating encryption keys. Their high sensitivity and accuracy are vital for the effectiveness and security of quantum communication.

- **Quantum sources of light**: Quantum cryptography requires sources that can emit single photons or entangled photon pairs. Technologies such as semiconductor quantum dots and parametric down-conversion are commonly used to produce these quantum states with high precision.

Current Global Implementations and Pilot Projects

- Quantum cryptography initiatives are underway across various countries and by multiple companies, each exploring different aspects of quantum communications:

    - **China**: Notably, China has made significant advancements, including launching the world's first quantum satellite, Micius, which facilitates intercontinental quantum-secured communications.

    - **Europe**: Several European nations are actively developing quantum communication networks. For instance, the UK's Quantum Communications Hub is working on extending the reach of secure quantum communications through various terrestrial and free-space methods.

    - **Private sector**: Companies like ID Quantique in Switzerland and QuintessenceLabs in Australia are pioneering commercial quantum cryptography solutions, focusing on integrating quantum technologies with existing security systems.

Challenges in Practical Deployment

- **Scale**: Scaling quantum cryptographic systems from small-scale experiments to widespread practical use is a significant challenge. Quantum systems require maintaining quantum coherence over long distances, which is currently limited.

- **Integration with existing infrastructure**: Integrating quantum cryptographic methods with the existing digital infrastructure poses technical challenges, as traditional and quantum systems operate on fundamentally different principles.

Future Directions

- **Potential for global quantum networks:** The vision of a global quantum network entails a world interconnected by quantum links that enable ultra-secure communications. Such networks would utilize satellites, ground stations, and possibly underwater quantum communication lines to achieve global reach.

- **Integration with existing cryptographic systems:** As quantum technologies evolve, a hybrid approach combining classical and quantum cryptographic techniques will likely be necessary to ensure a smooth transition. This integration will allow organizations to start leveraging quantum-resistant cryptography without discarding existing infrastructure.

- **Quantum repeaters:** Quantum repeaters will play a crucial role in overcoming the distance limitations of quantum communications. By enabling entanglement distribution over long distances, these devices could extend the reach of quantum networks, connecting cities and countries with secure quantum channels.

The development and implementation of quantum cryptography are still in the nascent stages but are rapidly advancing. As technologies mature and overcome current limitations, quantum cryptography could fundamentally transform the security landscape, offering unprecedented protection based on the laws of quantum mechanics.

Looking forward, the integration of quantum and classical systems, the expansion of global quantum networks, and the practical application of quantum repeaters will dictate the pace at which this exciting field evolves.

Thought Challenge: The Implications of Quantum Cryptography

Imagine you are the CEO of a global financial institution that handles sensitive data daily. With the advent of quantum cryptography becoming more accessible and potentially integrating into existing security systems, you are faced with the decision of whether to adopt this new technology.

- Consider the potential benefits and risks associated with transitioning to quantum cryptography. What are the primary considerations you should take into account? How might this shift affect your organization's security, operations, and competitive edge?

Reflection: Ethical Considerations in the Adoption of Quantum Cryptography

The advent of quantum cryptography introduces technological advancements and complex ethical dilemmas concerning accessibility and fairness. Quantum cryptographic technologies promise unprecedented security, potentially creating environments where data breaches and cyberattacks are things of the past. However, this advanced security raises significant questions about access and control:

- **Who gets access to quantum cryptographic technologies?** As these technologies develop, they may initially be expensive and complex to implement, likely making them accessible only to organizations or nations with significant resources. This disparity could lead to a security divide where only the wealthy or technologically advanced can afford the best protection, leaving smaller entities and developing countries more vulnerable to cyber threats.

- **Equity and inclusion:** If quantum cryptography becomes the gold standard for security, ensuring equitable access will be crucial. Without broad access, there could be an increase in inequality, not just within countries but also internationally. This situation demands a global conversation and cooperation to develop frameworks that promote inclusiveness and fairness in the deployment of quantum cryptographic solutions.

- **Regulatory challenges:** With great power comes great responsibility, and the powerful capabilities of quantum cryptography will necessitate thoughtful regulation to prevent misuse. Governing bodies will need to consider how these technologies are implemented, ensuring that they do not infringe on individual rights or enable oppressive surveillance practices.

- **Ethical deployment:** Beyond regulation, the ethical deployment of quantum cryptography involves considering how these technologies are used in practice. It's essential to foster a culture of ethical consideration among those developing and deploying quantum cryptographic technologies, ensuring that they are used to benefit society rather than create further divisions or harm.

As we think about the future of quantum cryptography, we must engage in open dialogues about the ethical implications of its deployment. This includes discussions on who controls these technologies, who has access to them, and how to prevent the creation of an uneven playing field in cybersecurity.

Possible Answers to the Thought Challenge: The Implications of Quantum Cryptography

Answer:

Benefits of Transitioning to Quantum Cryptography:

- **Enhanced security**: Quantum cryptography, particularly through QKD, offers theoretically unbreakable encryption, significantly boosting security against both current and future threats, including those from quantum computers.

- **Future-proofing**: Early adoption of quantum cryptography can provide a competitive edge by safeguarding against the evolving risk of quantum attacks and ensuring data remains secure as quantum technologies advance.

Risks and Considerations:

- **Cost and implementation**: The shift to quantum cryptography requires substantial initial investment and extensive modifications to existing systems. Financial and logistical aspects of setting up quantum-secure networks and training staff are critical considerations.

- **Technological maturity**: Quantum cryptography is still developing, and ongoing issues like key generation rates and secure distribution distances need to be resolved before widespread implementation.

- **Regulatory compliance**: Adoption of new technologies can attract regulatory scrutiny. Compliance with international standards and laws is essential, especially in heavily regulated sectors like finance.

Impact on Operations and Competitiveness:

- **Operational efficiency**: Integrating quantum cryptography must be smooth to prevent operational disruptions and maintain customer trust.
- **Market positioning**: Early adoption can enhance a company's reputation as a technological leader, potentially attracting customers who prioritize advanced security.

The decision to adopt quantum cryptography involves balancing the cutting-edge security benefits against the costs, technological maturity, and regulatory considerations. Fostering strategic discussions and thorough evaluations will be essential to making an informed decision that aligns with the institution's long-term goals and risk management strategies.

Chapter 9:

Quantum Mechanics in Everyday Life

Why does your coffee stay warm? Why is the sky blue? How does your phone manage to fit an entire world of information into your pocket? At first glance, these everyday phenomena might not seem to have much in common, but digging a little deeper, you'll find that quantum mechanics plays a crucial role in each of these scenarios. In this chapter, we'll explore how the seemingly abstract principles of quantum mechanics influence and drive many aspects of the world around us.

Throughout this book, we've uncovered the fascinating and often bewildering world of quantum physics, from the dual nature of light and matter to the entanglement phenomena Albert Einstein famously referred to as "spooky action at a distance." We've seen how these principles are poised to revolutionize technologies through quantum computing and cryptography. However, the influence of quantum mechanics isn't confined to high-tech laboratories and theoretical discussions; it permeates our everyday lives.

In this chapter, we discuss how quantum mechanics explains not just cutting-edge technology but also the fundamental processes that dictate the behavior of the very materials around us. From the molecules that make up your morning cup of coffee to the operation of the sun, quantum mechanics offers explanations where classical physics falls short.

So, let's reflect on these everyday interactions, armed with a deeper understanding of the quantum mechanics behind them, and perhaps you'll start to see the world a little differently—through the quantum lens.

Quantum Chemistry: The Science of Molecules

Quantum chemistry is the study of the quantum mechanical nature of atoms and molecules. It explains chemical properties and behaviors by applying the principles of quantum mechanics, such as wave-particle duality, quantization, and the uncertainty principle. These principles allow scientists to understand and predict the structure, stability, and reactions of molecules at a quantum level.

Molecular Bonding and Chemical Reactions

Electron Behavior in Atoms and Molecules

- At the heart of quantum chemistry is the electron, whose properties dictate how atoms bond and molecules behave. Electrons do not orbit the nucleus in defined paths as once thought in classical physics; instead, they exist in regions called orbitals, which are best described as probability distributions that indicate where an electron might be found at any given time.

Molecular Orbitals and Electron Configurations

- Molecular orbitals are formed by the combination of atomic orbitals when atoms bond to form molecules. These orbitals are the quantum states of the electrons in a molecule and can be bonding, antibonding, or non-bonding, each contributing differently to the molecule's stability and chemical properties.

- Electron configurations refer to the arrangement of electrons in an atom or molecule. Quantum chemistry uses principles like the Pauli exclusion principle and Hund's rule to explain how electrons fill orbitals, thus dictating the chemical and physical properties of the substance.

Bonding and Reactions

- Quantum mechanics explains chemical bonding through concepts like covalent bonds, where electrons are shared between atoms; ionic bonds, resulting from the electrostatic attraction between ions; and metallic bonds, involving a "sea" of delocalized electrons. These bonds and interactions are the result of the minimization of energy states, a key concept in quantum theory.

The textured spheres represent atomic orbitals. The overlapping of these spheres suggests the concept of orbital hybridization, which is crucial for the formation of covalent bonds. The lines interconnecting the spheres indicate electron pathways or interactions, highlighting how electrons are shared or transferred between atoms to form various chemical bonds.

Real-World Applications

- **Pharmaceutical development**: Quantum chemistry plays a crucial role in the pharmaceutical industry by helping to design and optimize drugs at the molecular level. By understanding how different molecular structures interact with biological targets, researchers can predict the efficacy and side effects of new drugs, speeding up the development process and improving drug performance.

- **Materials science**: In materials science, quantum chemistry is essential for discovering and designing materials with specific properties. For instance, superconductors, semiconductors, and nanomaterials are all developed with an understanding of the quantum mechanical interactions within materials. The ability to predict how changes at the molecular level affect macroscopic properties enables scientists to engineer materials with optimized electrical, mechanical, or thermal properties.

Thought Challenge: Quantum Chemistry in Everyday Products

Consider a common household item like a non-stick frying pan. The non-stick property of the pan is due to a coating often made from Teflon, which is a polymer consisting primarily of carbon and fluorine atoms. The effectiveness of Teflon and its characteristics, such as heat resistance and low friction, are deeply rooted in the interactions between its molecular structures.

- How does the application of quantum chemistry help explain why Teflon behaves the way it does in your kitchen? What does this suggest about the impact of quantum mechanics on everyday materials?

Quantum Biology: Quantum Effects in Nature

Quantum biology investigates how organisms leverage quantum phenomena to perform biological functions. Traditionally, biology and quantum mechanics have seemed worlds apart, with life's messy, wet environments appearing hostile to delicate quantum states. However, recent research suggests that life has evolved mechanisms to harness quantum effects for survival, offering a new perspective on the interplay between quantum physics and the living world (Alexander, 2023).

Quantum Effects in Biological Systems

Quantum Coherence in Protein Structures

- Quantum coherence, where particles are linked through their wave-like properties, is believed to play a role in the structure and function of proteins. Proteins are complex molecules essential to all biological systems, and their functions are dictated by their structure.
- Quantum coherence might explain how proteins achieve their highly efficient energy transfer and transformation, which classical physics struggles to explain. For instance, the precise mechanisms of enzyme action and protein folding might involve coherence at a quantum level, facilitating processes crucial for life.

Efficiency of Photosynthesis

- Photosynthesis, the process by which plants convert sunlight into chemical energy, is one of the most important biological processes on Earth. Recent studies suggest that quantum coherence allows for the highly efficient transfer of energy through light-harvesting complexes to reaction centers, where it is converted to chemical energy.

- Quantum coherence in the chlorophyll molecules helps explain how these transfers occur almost instantaneously and with minimal energy loss, enhancing the efficiency of photosynthesis beyond what classical models predicted.

Bird Navigation

- Many species of birds navigate using the Earth's magnetic field, a capability known as magnetoreception. One theory suggests that this ability is facilitated by a quantum phenomenon known as entanglement in a protein called cryptochrome, which is found in the retinas of migratory birds.

- According to this theory, light strikes the cryptochrome, triggering a chemical reaction that produces a pair of radicals (molecules with unpaired electrons). Quantum entanglement between these radicals' spins may allow birds to sense the Earth's magnetic field as a directional cue, enabling their remarkable navigational abilities.

Thought Challenge: Quantum Biology in Natural Processes

Consider the process of photosynthesis, a fundamental biological function that sustains plant life and, by extension, all life on Earth. This process involves plants converting light energy from the sun into chemical energy stored in glucose. Recent studies suggest that quantum coherence, a phenomenon where particles are connected over short timescales, may play a role in enhancing the efficiency of the energy transfer during photosynthesis.

- How does the concept of quantum coherence potentially explain the high efficiency of photosynthesis? What implications does this have for our understanding of other biological processes?

Quantum Sensors and Metrology: Precision and Innovation in Daily Life

Quantum sensors represent a significant breakthrough in measurement science, offering levels of precision and sensitivity that were previously unattainable. These devices exploit quantum phenomena such as superposition and entanglement to measure physical quantities with extraordinary accuracy.

What Are Quantum Sensors?

Quantum sensors utilize the inherent properties of quantum states to measure physical quantities like time, frequency, temperature, magnetic fields, and gravitational forces. These sensors operate on the principle that even the smallest changes in a physical quantity can alter a quantum system's state, which can be precisely measured.

For instance, atomic clocks, which are a type of quantum sensor, measure time by tracking the frequency of microwaves emitted by electrons changing energy levels within atoms.

How Quantum Sensors Work

Quantum sensors typically involve isolating and manipulating quantum particles such as photons, electrons, or atoms in controlled environments. Techniques such as laser cooling are often used to reduce the noise from thermal vibrations, enhancing the sensor's sensitivity and accuracy. The quantum properties of these particles, like their spin or polarization, are then used as a reference to detect changes in the environment that influence these properties.

Specific Examples of Quantum Sensors in Use Today

- **GPS technology:** One of the most well-known applications of quantum sensors is in the atomic clocks used in GPS satellites. These clocks utilize the precise oscillations of cesium or rubidium atoms to keep time with exceptional accuracy. The time measured by these atomic clocks is critical for determining a user's position on Earth, as even a microsecond's error can lead to significant errors in positioning.

- **Medical imaging:** Quantum sensors are also transforming medical imaging technologies. For example, devices based on nitrogen-vacancy centers in diamonds can detect the magnetic fields generated by neurons in the brain, allowing for non-invasive monitoring of brain activity. This technology could lead to better diagnostics for neurological conditions without the need for more intrusive methods.

Impact of Quantum Technologies

- **Unprecedented precision in measurements:** Quantum metrology, the science of measurement using quantum techniques, provides unprecedented precision, which is vital for many scientific and industrial applications. For example, in scientific research, precise measurements are crucial for testing the validity of theoretical models, especially in fields like particle physics and cosmology.

- **Implications for science and industry:** The heightened precision of quantum measurements can lead to innovations in various industries. For instance, more accurate sensors can significantly improve navigation systems, telecommunications, and geological surveys by detecting minute changes in environmental conditions. Additionally, quantum sensors could play a crucial role in emerging technologies like autonomous vehicles by enhancing the accuracy and reliability of obstacle detection systems.

Thought Challenge: Quantum Sensors and Everyday Precision

Imagine you are using a GPS device to navigate through a city. The GPS system's accuracy and reliability heavily depend on the precise timing signals from satellites equipped with atomic clocks. These atomic clocks are a type of quantum sensor that utilizes the principles of quantum mechanics to keep time with unprecedented accuracy.

- How does the application of quantum sensors in GPS technology affect daily life, particularly in navigation? What does this tell us about the broader impacts of quantum metrology?

Environmental Impact of Quantum Technologies

Finally, as we explore the expansive reach of quantum technologies into various aspects of daily life and industry, it's crucial to consider their environmental implications, such as the green footprint of these technologies, their energy efficiency, and potential environmental benefits, as well as the sustainability of the materials and methods used in their development.

Energy Efficiency of Quantum Technologies

Quantum computers, in theory, offer significant energy savings over classical supercomputers. Classical computers require exponential increases in power consumption to solve certain complex problems, which quantum computers might handle with drastically lower energy requirements due to their ability to process vast amounts of data simultaneously.

However, the current reality is that quantum devices, especially those requiring near-absolute zero temperatures, can be energy-intensive due to the cooling systems they necessitate. As technology progresses, the focus is on developing more energy-efficient quantum systems that could reduce overall power consumption while performing tasks unimaginable for classical computers.

Potential of Quantum Technologies to Monitor and Combat Environmental Issues

Quantum technologies have promising applications in environmental monitoring and protection. For example, quantum sensors can detect minute changes in environmental conditions, providing precise data for monitoring air quality, water pollution levels, and even the early signs of climate change.

Additionally, quantum geographic information systems (GIS) and remote sensing technologies could revolutionize how we manage natural disasters and track environmental degradation, leading to more effective conservation strategies.

Sustainability Considerations

- **Materials and methods used in quantum tech:** The sustainability of quantum technologies also hinges on the materials and production methods used. Many quantum computing components, such as those used in superconducting qubits, require rare or complex materials that can be challenging and environmentally taxing to mine and refine. There is a growing push within the industry to find more sustainable materials and improve the environmental footprint of the production processes. This includes reducing waste, recycling more components, and using greener energy sources during manufacturing.

- **Long-term sustainability:** As the adoption of quantum technologies grows, so does the responsibility to consider their long-term impact on the environment. Researchers and companies are beginning to address these issues proactively by designing systems that are not only effective but also sustainable. This involves ongoing research into how these technologies can be made more energy-efficient and less reliant on scarce resources.

Thought Challenge: Balancing Quantum Advancements With Environmental Sustainability

Imagine a world where quantum technologies have become mainstream, enhancing sectors like computing, communication, and medicine. As these technologies develop, they use materials and energy in ways different from traditional technologies. Quantum computers, for example, may require conditions near absolute zero temperature to function, potentially leading to high energy demands for cooling systems.

- Consider the potential environmental impacts of widespread quantum technology deployment. How can we balance the benefits of quantum advancements with the need for environmental sustainability? What strategies might be employed to mitigate potential negative impacts on the environment?

Reflection: Quantum Mechanics and the Fabric of Daily Life

Imagine your morning routine enhanced by quantum computing: your smart home system optimizes your schedule and household tasks based on predictive algorithms that make AI assistants seem primitive. Your coffee machine, synced to your biometrics, brews a cup perfectly tailored to your body's morning needs.

On your commute, traffic lights managed by quantum algorithms reduce congestion efficiently, adapting in real-time to traffic flow changes. The public transport systems, equipped with quantum sensors, run with optimal precision and safety, reducing delays and energy use.

In the workplace, quantum-enhanced virtual reality meetings could offer real-time language translation and much richer interaction nuances, making remote work more engaging and productive than ever before.

In your leisure time, perhaps you'll visit a park where quantum environmental sensors monitor and adjust conditions to maintain an ideal ecosystem. Or you might use a quantum fitness device that adjusts your workout in real time based on your body's quantum biofeedback, maximizing your health benefits.

While these enhancements will bring numerous benefits, our reflection must go beyond convenience and efficiency. As we familiarize ourselves with the presence of quantum technologies in our lives, the challenge will be to maintain a sense of community and personal connection in a world where technology has the potential to isolate us in bubbles of tailored experiences. How will we ensure that human interactions remain genuine and meaningful in a highly optimized, quantum-assisted world?

In anticipating the widespread adoption of quantum technologies, let's also commit to fostering discussions that ensure these advancements enhance not only our efficiency but also our well-being and societal bonds. As quantum technology becomes familiar, our responsibility grows—to guide its development thoughtfully and to craft a future where technology serves humanity, enhancing our lives without diminishing our values.

Possible Answers to the Thought Challenges

Quantum Chemistry in Everyday Products

Answer:

Quantum chemistry deeply influences everyday products, particularly in materials like Teflon. It explains Teflon's properties through electron configurations and molecular interactions at an atomic level. Teflon's strength derives from strong carbon-fluorine bonds, among the toughest in organic chemistry, explained by molecular orbitals formed from shared electrons. This contributes to Teflon's exceptional durability and heat resistance. Furthermore, Teflon's non-stick quality comes from its molecular structure, where fluorine atoms shield the carbon chain, repelling other substances. This interaction, crucial for understanding Teflon's properties, is clarified through quantum chemistry's analysis of molecular interactions.

Quantum Biology in Natural Processes

Answer:

Quantum coherence may significantly enhance the efficiency of natural processes like photosynthesis in plants. Instead of following a single pathway, absorbed energy from chlorophyll can simultaneously traverse multiple pathways due to the superposition principle. This allows energy to move through the chlorophyll network via the quickest and most efficient route, potentially explaining how plants achieve nearly 100% efficiency in converting solar energy into chemical energy. The implications extend beyond photosynthesis; similar quantum phenomena might influence other biological systems, such as avian navigation via Earth's magnetic field or the molecular detection mechanisms in human olfaction.

Quantum Sensors and Everyday Precision

Answer:

Quantum sensors, such as atomic clocks, significantly enhance GPS technology by providing extremely precise time measurements, crucial for the accuracy of GPS signals. Minor timing errors can lead to substantial navigation discrepancies. This precision is vital for applications ranging from everyday commuting to critical fields like aviation, military operations, and emergency services, ensuring safety and efficiency. Beyond navigation, quantum sensors improve other areas, such as medical imaging, enabling faster, higher-resolution MRIs, and environmental monitoring, where they detect subtle changes in atmospheric conditions or pollution levels, enhancing response capabilities to environmental challenges.

Balancing Quantum Advancements with Environmental Sustainability

Answer:

The integration of quantum technologies brings substantial advancements but also poses challenges to environmental sustainability. The key to balancing these aspects lies in proactive planning and the development of sustainable practices within the quantum technology sector.

- **Energy efficiency:** Quantum computers and their cooling systems consume significant energy. Potential solutions include:
  - developing energy-efficient quantum computation methods
  - improving cooling technology efficiency
  - utilizing renewable energy sources to power these technologies, reducing the carbon footprint
- **Sustainable materials:** Quantum devices often require rare or hazardous materials, posing sustainability and disposal challenges. Approaches to address this include:
  - developing recycling protocols for quantum components
  - researching alternatives that are less harmful and more abundant
- **Lifecycle assessment:** Conducting comprehensive lifecycle assessments can help identify environmental hotspots and intervention points across all stages—material extraction, device fabrication, usage, and end-of-life disposal.
- **Regulation and policy development:** Implementing robust regulations and developing policies that encourage environmentally responsible practices in the quantum technology field are essential. Incentives could include grants, subsidies, or tax breaks for companies that prioritize green practices.

Chapter 10:

Quantum Physics and Relativity—Bridging Theories

Think of a dance performance where the dancers, representing particles, move flawlessly in tune with the music, symbolizing the forces that govern them. On one side of the stage, dancers pirouette precisely, embodying the elegant and predictable world of relativity with its large-scale gravitational forces.

On the other, performers pop and lock in seemingly random bursts, illustrating the unpredictable quantum realm where particles exist in multiple states and interact in ways that defy classical intuition. The challenge, then, is for these two groups to come together in a single choreographed piece that captures the essence of both styles—this is the task of integrating quantum physics with relativity.

Throughout our discussions, we've ventured deep into the realms of quantum mechanics, examining its profound implications and the technologies it has inspired. We've also touched upon the grand scale of general relativity, where massive celestial bodies dictate the movement of the cosmos through the curvature of spacetime. However, these two foundational pillars of modern physics have historically operated as distinct performances, each in its own domain.

This chapter goes into the efforts to unify quantum mechanics and relativity, introducing advanced concepts like the Klein-Gordon and Dirac Equations, which address high speeds and minuscule scales. We'll also discuss the Standard Model of Particle Physics, which classifies fundamental particles and interactions but lacks integration with gravity. Additionally, we explore theories like Gravitational Quantum Mechanics and String Theory, which ambitiously attempt to harmonize the forces of gravity and quantum mechanics.

Our goal is not only to understand these theories but to appreciate their potential to fundamentally alter our understanding of the universe. Like a choreographer who blends different dance styles into a single performance, scientists are striving to blend quantum mechanics and relativity into a unified picture of how everything in the universe works—from the tiniest quarks to the vastest galaxies.

Special Relativity and Quantum Mechanics: Foundational Principles

Special relativity, a theory introduced by Albert Einstein in 1905, fundamentally altered our comprehension of the universe's structure and how we measure physical phenomena. By examining its core principles and their impact on quantum mechanics, particularly at velocities nearing the speed of light, we can grasp the profound adjustments required in the quantum domain.

Special relativity revolutionized our understanding of physics with two fundamental principles:

- **Constancy of the speed of light:** The speed of light in a vacuum is a universal constant at approximately 299,792 kilometers per second, unchanged by the observer's motion. This challenges classical notions of additive velocities and establishes a universal speed limit.

- **Relativity of simultaneity:** Time and simultaneity are relative. Events seen as simultaneous by one observer can occur at different times for another observer moving at a different speed.

This alters our understanding of time and space as interconnected dimensions.

Integration With Quantum Mechanics

Applying these principles to quantum mechanics, particularly for particles at near-light speeds, requires adjustments:

- **Relativistic mass:** As particles near light speed, their effective mass increases as viewed by an observer, influencing how quantum mechanics predicts their high-speed behavior.
- **Time dilation:** Time slows down for fast-moving objects relative to stationary ones, affecting the observed rates of decay and other time-sensitive processes in quantum systems.
- **Length contraction:** Objects moving close to light speed appear shorter in the direction of motion, which influences the analysis of their quantum wavefunctions and spatial distributions.

Thought Challenge: Visualizing Relativistic Effects in Quantum Mechanics

Imagine you are observing an experiment involving electrons being accelerated to near-light speeds. As these electrons approach the speed of light, their behavior is influenced by both quantum mechanics and the principles of special relativity.

- How would the concepts of relativistic mass, time dilation, and length contraction manifest in the behavior of these high-speed electrons? What implications would these effects have on the experimental results and our understanding of particle physics?

Quantum Field Theory (QFT) and Field Quantization Basics

QFT represents a pivotal integration of quantum mechanics with special relativity, which is crucial for understanding the fundamental interactions at subatomic scales and high energies. QFT redefines particles not as isolated entities but as excitations of fields that permeate the entire universe. These fields include electromagnetic, electron, and quark fields, among others, which are fundamental to QFT's framework.

- **Field excitations**: In QFT, particles such as electrons, photons, or quarks are essentially excited states of these fields. For example, an electron is considered an excitation of the electron field; similarly, a photon is an excitation of the electromagnetic field. This concept provides a unified view that accommodates the dual particle-wave nature observed in quantum experiments.

- **Particle creation and annihilation**: QFT equips physicists with tools to describe the creation and annihilation of particles through field interactions. During high-energy particle collisions, for instance, energy may convert into pairs like electrons and positrons, which emerge from the "excitation" of the vacuum state of their fields. These phenomena adhere to strict conservation laws, including energy and momentum.

- **Field quantization**: This process involves quantizing classical fields into operators with discrete energy levels, fundamentally altering their classical continuous nature. In this quantized framework, fields are treated as operators in Hilbert space, impacting quantum states with probabilistic outcomes akin to the uncertainty observed in particle positions and momenta.

- **Commutation relations**: Essential to QFT are the commutation relations between field operators, which ensure adherence to causality and locality principles. For fermions, such as electrons, these are anticommutation relations due to the Pauli exclusion principle, while bosons, like photons, follow commutation relations.

QFT underpins the Standard Model of particle physics, explaining the interactions among the strong, weak, and electromagnetic forces. Its predictions, such as the electron's magnetic moment and the Higgs boson's mass, have been confirmed with remarkable precision by experiments at facilities like the Large Hadron Collider (Anastasiou, 2020).

The Klein-Gordon Equation

The Klein-Gordon equation, developed in the 1920s, was a pioneering attempt to incorporate relativity into quantum mechanics. It focuses on high-energy conditions where relativistic effects are pronounced. This equation integrates the relativistic energy-momentum relation $E^2 = p^2c^2+m^2c4$, applying it within the quantum framework to describe particles in extreme conditions.

Primarily relevant to bosonic particles that don't adhere to the Pauli Exclusion Principle, the Klein-Gordon equation addresses mesons and similar force-carrying particles. Unlike fermions, which are governed by exclusionary rules preventing two identical particles from occupying the same state, bosons can occupy the same state in multiples, which is crucial for explaining strong nuclear forces in atomic nuclei.

Mathematically, the Klein-Gordon equation is expressed as:

$$(\Box + m^2c^2/\hbar^2)\psi = 0$$

Where:

- $\Box$ denotes the d'Alembert operator, defined in Minkowski spacetime as $\Box = \partial^2/\partial t^2 - \nabla^2$.
- ψ is the field representing the particle.

This operator effectively incorporates the second derivative with respect to time, aligning the equation with the requirements of special relativity.

However, initial formulations of the Klein-Gordon equation faced issues like predicting negative probabilities, which conflicted with physical interpretations of probabilities as non-negative. These problems stemmed from the equation's second-order time derivative, unlike the first-order Schrödinger equation.

Over time, advancements in Quantum Field Theory (QFT) refined its application, shifting interpretation from individual particles to fields, thus addressing the earlier criticisms by treating ψ\psiψ as a field operator, not just a probability amplitude. This reinterpretation allowed the equation to describe particle and antiparticle excitations within quantum fields effectively, confirming its essential role in modern particle physics for exploring the properties and interactions of scalar and pseudoscalar bosons, thereby enriching the theoretical foundation for further advances in the field (Anastasiou, 2020).

Thought Challenge: Understanding Mesons With the Klein-Gordon Equation

You're reviewing results from a recent particle collider experiment where various mesons were detected. These mesons, which are important for studying the strong nuclear force, are often analyzed using the Klein-Gordon equation.

- How does the Klein-Gordon equation facilitate predictions about the behaviors and properties of mesons in these experiments? What does this tell us about the fundamental forces at play?

The Dirac Equation

The Dirac Equation, formulated by Paul Dirac, significantly advanced the integration of quantum mechanics with special relativity, focusing particularly on fermions like electrons and positrons with a spin of ½. This equation was pivotal for accommodating the intrinsic spin of particles, addressing limitations of the prior Klein-Gordon equation, which failed to manage spin-statistics and yielded non-physical negative probabilities.

The equation is expressed through the use of gamma matrices, which handle the anticommutation essential for fermion behavior:

$$(i\gamma^\mu \partial_\mu - m)\psi = 0$$

Here:

- γ^μ are the gamma matrices.
- ∂_μ represents the partial derivatives for spacetime coordinates.
- m is the mass of the fermion.
- ψ is a spinor field representing the quantum state of the particle.

Key Implications and Discoveries

- **Antiparticles:** One of the revolutionary predictions of the Dirac Equation was the existence of antiparticles, asserting that each particle has a counterpart with an opposite charge. This concept was validated by the discovery of the positron by Carl Anderson in 1932, reinforcing the equation's significance in particle physics and supporting technologies like PET scans.
- **Spin and quantum statistics:** The Dirac Equation inherently includes fermion spin-½ characteristics, respecting the Pauli Exclusion Principle that prevents identical fermions from occupying the same quantum state. This principle is fundamental in understanding atomic and molecular structures, influencing the development of semiconductors and other electronic technologies.
- **Relativistic effects:** The equation incorporates special relativity, which is crucial for predicting how particles behave at speeds approaching that of light. Such relativistic considerations are

vital for interpreting phenomena in high-energy environments like particle accelerators at CERN, where particles exhibit altered mass, time dilation, and length contraction due to their high velocities.

Thought Challenge: The Dirac Equation and Everyday Electronics

Next time you use your smartphone or any electronic device, consider that the functionality of these devices hinges significantly on the principles described by the Dirac Equation.

- How does the Dirac Equation, which describes the behavior of electrons at relativistic speeds, impact the development and functionality of modern electronic devices?

The Standard Model of Particle Physics: A Particle Overview

Another implication of those theories that incorporate both quantum mechanics and relativity is the Standard Model. This is an incredibly successful scientific theory that provides a comprehensive framework for understanding the fundamental particles and forces in the universe.

It classifies all known subatomic particles and describes three of the four fundamental forces—electromagnetism, strong nuclear force, and weak nuclear force—leaving out gravity, which is currently described by general relativity.

Fundamental Particles

The Standard Model categorizes the basic building blocks of matter and energy into two main groups: fermions and bosons.

- **Fermions**: These particles constitute all the matter in the universe. Fermions are divided into quarks and leptons.
    - **Quarks** make up protons and neutrons, which in turn make up the nuclei of atoms. There are six types (flavors) of quarks: up, down, charm, strange, top, and bottom.
    - **Leptons** include electrons, muons, tau particles, and their corresponding neutrinos. Each lepton flavor is associated with a specific type of neutrino.
- **Bosons**: These particles act as force carriers that mediate the fundamental forces of nature.
    - **Photon** is the carrier of electromagnetic force.
    - **Gluons** are responsible for the strong force that binds quarks together inside protons, neutrons, and other particles.
    - **W and Z bosons** mediate the weak force, which is responsible for radioactive decay.
    - **Higgs boson** is a unique scalar particle that gives mass to other particles through the Higgs field interaction.

Fundamental Forces

The Standard Model explains how forces act between fundamental particles through the exchange of bosons:

- **Electromagnetic force**: Mediated by photons, it acts between electrically charged particles.
- **Strong nuclear force**: The strongest force, mediated by gluons, it acts between quarks.
- **Weak nuclear force**: Mediated by W and Z bosons, it causes nuclear processes like beta decay.

The Higgs Mechanism

Central to the Standard Model is the Higgs mechanism, which explains how particles acquire mass. The Higgs field permeates all space, and particles interact with this field to varying degrees; the stronger the interaction, the greater the mass of the particle. The Higgs boson, discovered in 2012 at CERN, is a manifestation of quantum fluctuations in the Higgs field.

Gravitational Quantum Mechanics: The Quest for Unity

The Standard Model of Particle Physics effectively describes three of the four fundamental forces but omits gravity. The quest to unite general relativity (gravity's theory) with quantum mechanics is known as gravitational quantum mechanics, aiming for a unified theory that marries quantum properties with gravitational effects.

Current Theories and Developments

- **Quantum gravity:** This field attempts to describe gravity through quantum mechanics principles, a challenging task due to the fundamental differences between general relativity (cosmological scales) and quantum mechanics (subatomic scales).

- **String theory:** A leading theory proposing that particles are one-dimensional "strings" whose vibrations correspond to different particles, suggesting a unified model for all particles and forces, including gravity.

- **Loop quantum gravity (LQG):** LQG proposes that spacetime itself is quantized, made up of discrete chunks, and works within the known four dimensions, offering a potential framework to integrate quantum mechanics with Einstein's theories.

- **M-theory:** An extension of string theory, suggesting that strings in ten dimensions can appear as membranes in eleven dimensions, representing a more comprehensive unification approach, albeit highly speculative.

These theories represent the forefront of theoretical physics in combining gravity with quantum mechanics, each offering unique insights and challenges in bridging these foundational theories.

Reflection: How a Unified Theory Might Change Our View of the Universe

Currently, our best model for particle physics, the Standard Model, effectively describes three fundamental forces but leaves out gravity, which is governed by general relativity. Bridging this gap could revolutionize our understanding of the universe with:

- **A more complete understanding of the universe:** A unified theory could provide a seamless explanation that connects the behavior of the smallest elements of our universe with the vast expanses of cosmic phenomena. This would fill in the current gaps in our understanding and potentially explain mysterious aspects of our universe, like dark matter and dark energy.

- **Predicting new phenomena:** Just as relativity predicted the bending of light and quantum mechanics predicted the existence of antimatter, a unified theory might predict new phenomena or provide solutions to unresolved puzzles in physics. This could include new insights into the nature of black holes or even the origins of the universe itself.

How do you feel about the potential transformations a unified theory could bring? Are you excited about the new insights and technologies it could unlock, fundamentally changing our understanding of the universe and the fabric of our everyday lives?

Possible Answers to the Thought Challenges

Visualizing Relativistic Effects in Quantum Mechanics

Answer:

- **Relativistic mass increase**: As electrons approach the speed of light, their mass effectively increases, requiring more energy to accelerate. This alters their interaction dynamics and collision outcomes in particle experiments.

- **Time dilation**: Time dilation causes internal processes such as decay rates to appear slower for moving electrons compared to stationary ones. This impacts measurements of particle lifespans and decay processes in experiments.

- **Length contraction**: Electrons moving at relativistic speeds undergo length contraction, affecting their wavefunctions and, consequently, the outcomes of experiments like interference patterns in double-slit setups.

Understanding Mesons with the Klein-Gordon Equation

Answer:

The Klein-Gordon equation is crucial for modeling mesons because it predicts their mass, energy levels, and decay properties. These predictions are vital for identifying different types of mesons in experiments, helping physicists understand the strong nuclear force and the fundamental interactions of particles. This application of the equation shows how theoretical physics underpins practical experimental analysis, enhancing our grasp of particle behavior and interactions.

The Dirac Equation and Everyday Electronics

Answer:

The Dirac Equation explains the properties of electrons, including their spin and the presence of antimatter (positrons), which are fundamental for developing semiconductor technology used in all modern electronics. The spin properties of electrons, governed by the Dirac Equation, are crucial for the behavior of materials under different electrical conditions, influencing how electrons move through a semiconductor. This understanding is pivotal for designing more efficient transistors, which are the building blocks of integrated circuits found in everything from computers to smartphones.

Chapter 11:

Quantum Paradoxes and Thought Experiments

Life is full of paradoxes. Consider the simple act of deciding what to wear based on the weather forecast. If the forecast predicts rain and you carry an umbrella, the day might stay sunny, and you'll wonder if your preparedness somehow warded off the rain. While not a direct comparison, this reflects a similar conundrum faced in quantum mechanics, where the act of observation can seemingly influence the outcome.

This chapter explores the intriguing realm of quantum paradoxes, each one presenting a unique challenge to our classical understanding of physics. We'll explore concepts that might initially sound as perplexing as the weather scenario but are crucial for grasping the subtleties of quantum theory. From the famous EPR Paradox, which stretches our notions of connection and information transfer, to the mind-bending implications of the Delayed Choice Quantum Eraser, these paradoxes force us to question the very fabric of reality.

Quantum paradoxes often leave even seasoned physicists puzzled, reflecting the mysterious and often counterintuitive nature of the quantum world. They challenge our everyday experiences with their non-intuitive outcomes and push the boundaries of what we consider possible. So don't worry if you don't grasp everything at first—these concepts are meant to provoke thought and deepen our understanding of the universe, one paradox at a time.

The EPR Paradox: Revisiting Quantum Entanglement

Previously, when discussing entanglement and nonlocality and Bell's Theorem, we touched on the foundational aspects of the Einstein-Podolsky-Rosen (EPR) Paradox. Introduced by Albert Einstein, Boris Podolsky, and Nathan Rosen in 1935, remember how this paradox challenges the fundamental notions of quantum states and locality. Einstein and his colleagues presented a thought experiment where two particles interact and become entangled, meaning the state of one particle instantly correlates with its partner, regardless of the distance between them.

They posited that if one particle's position or momentum is measured, the corresponding property of the second particle is instantly known, suggesting a level of predetermined behavior not accounted for by quantum mechanics. This led them to propose that quantum mechanics might be incomplete, possibly missing "hidden variables" that predetermine these outcomes.

The EPR Paradox spurred further scrutiny and development in quantum theory, influencing John Bell's Theorem, which tests the paradox through empirical means. Bell's inequalities assess whether quantum mechanics or hidden variables more accurately predict reality.

Experiments, particularly those by Alain Aspect in the 1980s, have mostly validated quantum mechanics, supporting the inherent entanglement and nonlocality, and challenging classical ideas of causality. This continued affirmation of quantum predictions emphasizes the non-classical underpinnings of the theory, reshaping our understanding of quantum interactions.

Many-Worlds vs. Copenhagen Interpretations

The implications of the EPR Paradox extend beyond the mere mechanics of quantum physics, influencing profound philosophical debates about the interpretation of quantum mechanics itself. Two of the most prominent interpretations that arise in this context are the Many-Worlds Interpretation and the Copenhagen Interpretation, each offering a distinct perspective on the outcomes and fundamental nature of quantum events, like those described by the EPR Paradox.

Many-Worlds Interpretation

The Many-Worlds Interpretation, proposed by Hugh Everett in 1957, suggests that all possible outcomes of quantum measurements are physically realized, each in a different "world" or branch of the universe. According to this interpretation, the universe is constantly splitting into a multitude of parallel realities, each representing a different possible outcome of every quantum event (Ball, 2018).

In the context of the EPR Paradox, the measurement of a particle's state (like spin or position) in one part of the universe immediately results in the splitting of the universe into branches where each possible outcome of these measurements exists simultaneously but in different worlds.

This interpretation removes the need for a wave function collapse, which is a key component of the Copenhagen Interpretation, offering a deterministic and arguably simpler understanding of quantum mechanics, as it does not require observers to play a special role.

Copenhagen Interpretation

The Copenhagen Interpretation, developed mainly by Niels Bohr and Werner Heisenberg in the 1920s, remains one of the oldest and most widely taught interpretations. It posits that a quantum system remains in a superposition of states until it is observed, at which point the wave function collapses to a single outcome (Ball, 2018). Under this interpretation, quantum mechanics does not provide a description of an objective reality but only probabilities of observing various outcomes.

Regarding the EPR Paradox, the Copenhagen Interpretation would suggest that the entangled particles do not have specific attributes like position or momentum until one of the particles is measured. The act of measuring one particle instantaneously influences the state of the other, no matter the distance, leading to the collapse of the wave function for both particles simultaneously.

The Debate

The contrast between these interpretations highlights fundamental philosophical questions about reality and measurement in quantum mechanics. The Many-Worlds Interpretation offers a deterministic universe where every possibility is realized, leading to a potentially infinite number of universes.

In contrast, the Copenhagen Interpretation embraces a probabilistic universe where outcomes are only determined at the moment of measurement, emphasizing the role of observation in shaping reality.

Both interpretations attempt to address the nonlocality and entanglement highlighted by the EPR Paradox, yet they lead to vastly different conclusions about the nature of reality and the role of the observer. The debate between these interpretations continues to be a central theme in the foundations of quantum theory, as it challenges our understanding of the universe and the nature of existence itself.

Thought Challenge: Understanding Many-Worlds vs. Copenhagen Interpretations

You're watching a football game where a player shoots towards the goal. The ball either goes in or misses, but you only find out when the camera follows its trajectory. Now, apply this to quantum mechanics: in the Copenhagen interpretation, the ball's outcome is not determined until observed, whereas, in the Many-Worlds interpretation, both outcomes occur but in different universes.

- How do these interpretations differ in explaining the outcome of a quantum event, such as a particle's position measurement? What does this tell us about the nature of reality according to each interpretation?

The Quantum Eraser Experiment

The Quantum Eraser experiment is a fascinating demonstration of quantum mechanics that further deepens the mysteries surrounding observation and measurement, building on concepts like those seen in the EPR Paradox.

It elaborates on the famous double-slit experiment, where particles (such as photons or electrons) are fired at a screen through two slits. When unobserved, particles exhibit wave-like behavior, creating an interference pattern on the screen that indicates they have passed through both slits simultaneously — a phenomenon known as superposition.

However, if a measurement device is used to determine through which slit a particle passes, the particle exhibits particle-like behavior, and the interference pattern disappears, indicating that each particle goes through one slit or the other, but not both.

The Eraser Setup

In the Quantum Eraser experiment, an additional layer is added: after the which-path information (determining through which slit the particle passed) is obtained, another device, the "eraser," can modify or "erase" this information. Remarkably, if the which-path information is erased, the interference pattern reappears, suggesting that the particles revert to a state of superposition (Kim & Ham, 2023).

The Quantum Eraser experiment encapsulates a quantum paradox where the very act of observation and the timing of this observation can seemingly alter the past state of a particle. This challenges the classical ideas of fixed causality and objective reality, suggesting instead that at the quantum level, reality is a complex interplay of probabilities that are not settled until observed. It opens profound discussions about the nature of time, reality, and the role of the observer in shaping both.

Thought Challenge: The Quantum Eraser Experiment

You take a photo with your camera, but before you can look at it, you accidentally delete it. Usually, you'd assume the image is gone forever. However, the Quantum Eraser experiment suggests something more intriguing in the quantum world: Even after an event has occurred, "deleting" the information can change the outcome that was seemingly already decided.

- How does the Quantum Eraser experiment challenge the conventional idea of actions and consequences in the physical world?

Wigner's Friend Paradox: A Deeper Reflection Into Observer Effect

The Wigner's Friend paradox, first proposed by physicist Eugene Wigner, expands on the role of the observer in determining a quantum system's state. This thought experiment pushes the boundaries of the observer effect, highlighting the subjective nature of reality in quantum mechanics and challenging our understanding of objective reality.

The Wigner's Friend Setup

The experiment begins with a basic quantum scenario similar to Schrödinger's Cat thought experiment, involving a system that can be in a superposition of two states. Wigner's friend, who is inside a closed laboratory, measures this system. According to the principles of quantum mechanics, the act of measuring should collapse the system into one of the two definite states.

However, from outside the laboratory, Wigner himself does not know the outcome of his friend's measurement. For Wigner, both the system and his friend (who has observed the system) are together in a superposed state. It is only when Wigner opens the laboratory and asks his friend about the outcome that, according to the Copenhagen interpretation, the superposition collapses for Wigner (Crease, 2024).

The Core of the Paradox

- **Multiple observers**: The paradox arises because there are two observers: Wigner's friend, who makes an observation and collapses the quantum state from his perspective, and Wigner, who considers both the friend and the system still in superposition until he checks the result. This setup leads to conflicting accounts of reality based on the observer's interaction with the system.

- **Reality is observer-dependent**: If quantum mechanics is correct, then reality can differ between observers, even when they are describing the same event. This conflicts with the classical view of an objective, observer-independent reality.

- **Implications on consciousness and measurement**: This thought experiment also touches on the implications of consciousness in the measurement process. It questions whether consciousness is required to collapse a quantum state or if any interaction constituting a measurement would suffice, delving into the depths of what constitutes an 'observer.'

Thought Challenge: Wigner's Friend Paradox

Consider a scenario where you and a friend watch a magic trick involving a card that is both red and blue simultaneously, but when either of you guesses the color, it becomes the color guessed. You think the card is red, and when you check, it's red. Your friend, however, believes it's blue, and upon checking, it appears blue to them.

- How does Wigner's Friend Paradox reflect similar complexities in the understanding of quantum events, and what does this imply about the objectivity of observations?

Reflection: The Nature of Reality and Quantum Paradoxes

In exploring the perplexing world of quantum paradoxes, we enter into scenarios that seem as bewildering as trying to recall a memory that might not have happened exactly as we remember. Just as our memories can play tricks on us, making us unsure if an event occurred one way or another, quantum paradoxes challenge our understanding of what's truly happening at the fundamental levels of reality.

Quantum mechanics, through experiments like the Quantum Eraser or philosophical quandaries like Wigner's Friend, reveals a universe that is far less straightforward than our everyday experiences might suggest. These paradoxes illustrate that reality, at the quantum level, may not be a single, fixed sequence of events but a plethora of possibilities that depend heavily on when and how they are observed.

This chapter has shown us that reality might not only be stranger than we suppose but stranger than we *can* suppose. Like misremembered details of a story, each observation in the quantum realm can alter the narrative of reality, suggesting a universe that is dynamic and mutable, influenced by our interactions with it.

As we ponder these ideas, consider how this fluid concept of reality influences our understanding of the cosmos and the very essence of personal experience. Could the memories we trust and the reality we observe be as mutable and open to interpretation as the outcomes of these quantum experiments? What might this mean for our perception of truth and existence?

Possible Answers to the Thought Challenges

Understanding Many-Worlds vs. Copenhagen Interpretations

Answer:

The Copenhagen interpretation of quantum mechanics posits that quantum events remain in all possible states until observed, at which point they "collapse" into one outcome. This suggests a probabilistic reality influenced by observation. In contrast, the Many-Worlds interpretation asserts that all possible outcomes of a quantum event do occur but in separate, branching universes.

According to this view, every possible scenario happens in some version of reality, leading to a multiverse where every conceivable event unfolds in parallel. These interpretations present fundamental philosophical questions about the nature of reality and our role in shaping or merely observing it.

The Quantum Eraser Experiment

Answer:

In the Quantum Eraser experiment, classical physics' notion of fixed outcomes is challenged. Normally, recording which path information in a double-slit experiment causes particle-like behavior, eliminating the interference pattern. However, if this information is later erased, the wave-like interference pattern reappears, suggesting that future actions (erasing the information) can influence past events (the particle's path).

This phenomenon indicates that in the quantum realm, events remain in a superposition of all possible outcomes until all possibilities are observed or information is irreversibly recorded. This contrasts with the classical idea that outcomes are definite and immutable, introducing the possibility that outcomes can be altered by subsequent actions.

Wigner's Friend Paradox

Answer:

Wigner's Friend Paradox introduces a unique aspect of quantum mechanics involving observer-dependent reality. In the scenario, a quantum system observed by Wigner's friend inside a laboratory results in a definite outcome. However, to Wigner, outside the lab, both the system and the observation itself remain in a superposition until he verifies them. This paradox suggests that reality may be subjective, varying depending on the observer.

Much like differing perceptions of a card's color in a magic trick, a quantum particle can exist in multiple states simultaneously until observed. This challenges the classical view of a single, objective reality, proposing instead that different observers might experience and report different realities, thus questioning the universality of objective reality in quantum contexts.

Chapter 12:

The Future of Quantum Physics—What Lies Ahead

Throughout this book, we've journeyed through the fascinating landscape of quantum physics, uncovering its complex theories, groundbreaking experiments, and the deep philosophical questions it poses. We've explored how quantum mechanics challenges our classical views of reality and examined its ethical implications and the transformative potential it holds for the future.

As we've discovered, quantum physics doesn't just reshape our understanding of particles and forces; it offers profound insights into the fabric of the universe itself. But the journey doesn't end here. This chapter will take us even deeper into the quantum realm to explore emerging theories about the universe's origins, the nature of entropy, the direction of time, and even the possibility of a multiverse.

These topics stretch the boundaries of quantum mechanics into the territories of cosmology and theoretical physics. They prompt us to ask even bigger questions: How did the universe begin? What will its fate be? And what does quantum mechanics tell us about the arrow of time?

The Quantum Nature of Spacetime

The concept of spacetime, a seamless blend of the three dimensions of space with the dimension of time into a single four-dimensional continuum, is a cornerstone of Einstein's theory of general relativity. However, as we enter into the quantum realm, the traditional understanding of spacetime begins to unravel, revealing a landscape where the fabric of spacetime itself exhibits quantum properties. This shift is essential for advancing our understanding of the universe's most fundamental aspects, including black holes, the Big Bang, and the underlying structure of the cosmos.

Quantum Fluctuations in Spacetime

In classical physics, spacetime is a smooth, continuous fabric. Yet, quantum theory introduces the concept of quantum fluctuations—temporary changes in the amount of energy in a point in space, which suggests that spacetime itself might be composed of discrete, quantized units at the smallest scales. This quantum granularity implies that spacetime could have a 'foamy' structure when viewed at the Planck scale (around 10^−35 meters), where the effects of quantum gravity become significant.

Loop Quantum Gravity and Spacetime

One of the leading theories that explores the quantum nature of spacetime is Loop Quantum Gravity (LQG). LQG proposes that spacetime has a looped, network-like structure at the quantum level composed of tiny, discrete loops (Vyas & Joshi, 2021). These loops are not just theoretical constructs but are quantized threads of spacetime, each one with a size on the order of the Planck length. This theory stands as a potential way to merge quantum mechanics with general relativity, providing a framework that includes gravity in the quantum description of the universe.

Black Hole Thermodynamics

The interplay between black hole physics and quantum mechanics, particularly through the lens of quantum spacetime, has led to significant revisions in our understanding of black holes. Historically viewed as ultimate endpoints where all matter and information irretrievably vanish, black holes, under the quantum microscope, reveal more nuanced behaviors, suggesting they may not completely erase information as once believed.

Quantum spacetime posits that the fabric of spacetime is composed of discrete, quantized units. This granular nature fundamentally alters how we perceive phenomena occurring at the event horizons of black holes. One of the pivotal insights from applying quantum mechanics to black holes is the realization that the quantum effects inherent to spacetime can influence black hole dynamics, particularly how information is handled.

Hawking Radiation: A Pathway for Information

Stephen Hawking's groundbreaking work introduced the concept of Hawking radiation, which proposes that particle-antiparticle pairs can spontaneously form near the event horizon of a black hole. In classical physics, these pairs would annihilate each other almost instantaneously; however, in the presence of a black hole, one particle may fall into the black hole while the other escapes. This process leads to the black hole losing mass over time, theoretically until it evaporates completely (Siegel, 2023).

Hawking radiation is significant because it suggests a mechanism by which information could escape from a black hole. The particles that escape carry away information, countering the longstanding belief that all information entering a black hole is lost forever. This process aligns with the principles of quantum mechanics, which maintain that information cannot be destroyed.

Information Paradox and Quantum Spacetime

The concept of information escaping from black holes via Hawking radiation ties directly into the broader "information paradox," which questions how information can seemingly disappear in a black hole if physical laws dictate that it should be conserved. Quantum spacetime offers a framework within which this paradox can be explored further. If spacetime is quantized, the very nature of how events unfold at the event horizon—where quantum gravitational effects dominate—may be different from traditional predictions.

The study of how quantum spacetime might resolve or reformulate the information paradox is ongoing. Researchers are exploring whether the discrete structure of spacetime could alter the behavior of black holes in such a way that information is not lost but rather transformed and emitted back into the universe through complex quantum processes (Weinstein, 2023).

Thought Challenge: Understanding Hawking Radiation

Consider how a sponge slowly releases absorbed water when it's squeezed. This analogy can help us grasp the concept of Hawking radiation, where a black hole, influenced by quantum effects near its boundary, emits particles in a seemingly unlikely process.

- How does the concept of Hawking radiation challenge the classical view that nothing can escape a black hole?

Big Bang and Cosmology: Quantum Insights into the Universe's Origins

The incorporation of quantum mechanics into our understanding of spacetime has revolutionary implications for cosmology, particularly concerning theories about the origin of the universe, such as the Big Bang.

Classical descriptions of the Big Bang suggest a singular origin, where the universe's density and temperature reached infinite values. However, the notion of quantum spacetime introduces a new perspective that challenges and potentially refines this classical model.

The Big Bang Singularity

In classical cosmology, the Big Bang is often described as a singularity—an initial point where the physical theories we rely on no longer hold, and quantities that describe the state of the universe, like density and temperature, become infinite. This singularity poses significant theoretical challenges, as it suggests limits beyond which physics cannot explain conditions or predict outcomes.

Quantum theories of gravity, including those positing a quantized spacetime, suggest that these singularities may not exist in the way classical theories propose. Instead, if spacetime itself is made up of discrete, quantized units, the very fabric of the universe at its smallest scales is not infinitely divisible (Oppenheim, 2023). This quantization implies that there is a fundamental limit to the smallness of spacetime, smoothing out what appears as singularities in classical theories.

Implications for the Big Bang

Applying the concept of quantized spacetime to the Big Bang leads to several profound implications:

- **A smoother beginning**: Quantum corrections to the classical theory of the Big Bang suggest that the "singularity" might instead have been an extremely dense and hot state but not truly infinite in density or temperature. This scenario proposes a more smoothed-out origin, where the universe's initial state is finite and describable within a quantum framework.

- **Avoiding infinite quantities**: By introducing a minimum possible scale of spacetime, quantum spacetime prevents physical quantities from reaching infinity. This could mean that the universe began from a state that, while extraordinarily dense and hot, was still finite and governed by laws that consistently apply through time without needing new physics at infinite values.

- **Pre-Big Bang scenarios**: Quantum models of spacetime open the door to new theories about what preceded the Big Bang. These models include the possibility of a bouncing universe, where the Big Bang was preceded by a Big Crunch of a previous universe or scenarios where our universe is one bubble in a larger multiverse.

Testing and Observations

Current cosmological observations, such as measurements of the cosmic microwave background radiation and the distribution of galaxies, are consistent with a Big Bang origin. However, these observations do not yet provide detailed information about the state of the universe at its very beginning. Future observations and theoretical developments in quantum gravity may offer more direct insights into whether spacetime is quantized and how such quantization would affect the initial conditions of the universe.

Thought Challenge: Quantum Insights Into the Big Bang

Think about watching a fireworks show. From the ground, we see a dazzling burst that seems to emerge from a single point, expanding rapidly into an array of colors and patterns. Now, apply this to the universe's beginnings in the Big Bang, traditionally viewed as a singular explosion from an infinitely small point.

- How might the idea of quantized spacetime change our understanding of the Big Bang, similar to discovering that the fireworks are actually a series of many smaller, controlled bursts rather than a single explosion?

Entropy, Time's Arrow, and the Multiverse: Quantum Perspectives on Universal Phenomena

The concepts of entropy, time's direction (often referred to as Time's Arrow), and the possibility of a multiverse are central to understanding how the universe functions at both macroscopic and microscopic levels. Quantum mechanics, with its insights into the fundamental aspects of nature, offers unique perspectives on these phenomena, challenging and extending our classical views.

Entropy and Quantum Mechanics

Entropy, traditionally associated with the degree of disorder or randomness in a system, is a critical concept in both thermodynamics and information theory. In the quantum realm, entropy is not just a measure of disorder but also of uncertainty or the information content inherent in a quantum state. Quantum entropy can be seen in the entanglement of particles, where the entangled state's entropy can increase or decrease based on the observer's knowledge of the system.

- **Quantum entanglement and entropy**: Quantum entanglement introduces a type of entropy that reflects correlations between entangled particles that are not present in classical systems. This quantum entropy can be higher or lower depending on whether measurements collapse the wave functions of entangled particles, thereby changing the information we have about the system.

- **Entropy and quantum decoherence**: Decoherence, which describes how quantum systems interact with their environment and lose their quantum behavior, plays a significant role in understanding the quantum basis of entropy. Decoherence increases entropy as the system transitions from a well-defined quantum state to a mixed state due to interactions with the environment.

Time's Arrow and Quantum Mechanics

Time's arrow refers to the one-way direction of time from past to future, closely associated with the second law of thermodynamics, which states that entropy tends to increase over time. In classical mechanics, the laws are time-symmetric, meaning they do not inherently define a direction of time. However, quantum mechanics introduces elements that can give a direction to time's arrow through the measurement process.

- **Measurement and time's arrow**: The act of measurement in quantum mechanics can create a definite state out of a superposition, effectively increasing entropy and giving a direction to time. This is seen as aligning with Time's Arrow, as the quantum state's collapse appears irreversible, mirroring the thermodynamic arrow of time.

The Multiverse and Quantum Mechanics

The concept of the multiverse arises naturally from quantum mechanics, particularly from the Many-Worlds interpretation. This interpretation suggests that all possible outcomes of quantum measurements actually occur, each in its own separate universe, thus comprising a vast multiverse.

- **Multiverse and quantum decisions**: Every decision and measurement at the quantum level could potentially spawn new universes, each representing different outcomes of quantum processes. This ties into the idea of quantum superpositions existing in multiple states simultaneously until measured.

- **Multiverse implications for entropy and time**: In a multiverse, entropy can vary across different universes depending on their states and histories. Moreover, the direction of time could also differ, with some theoretical models suggesting universes where time could run backward or have different properties based on their quantum states.

Reflection: The Multiverse and Everyday Decisions

Just as the multiverse suggests countless possibilities springing from every quantum event, each decision we make—from mundane choices like what to wear to life-changing decisions like career moves—can lead us down vastly different paths. Though we don't see these paths as tangible realities, considering them can help us appreciate the richness of options available and the impact of our decisions.

The multiverse theory could be likened to watching a movie like *Mr. Nobody*, where every decision the protagonist faces leads to a branching of realities, each with its own narrative and outcome. While we might not experience these alternate realities, the awareness of their potential existence can affect how we value our choices and perceive their consequences.

- How might embracing a multiverse perspective alter your approach to decision-making? If every choice potentially creates a new "branch" in your personal life story, does this make your daily decisions feel more significant or daunting?

- Could this perspective encourage you to take more risks or explore more diverse paths, knowing that, in some sense, every possibility is realized somewhere?

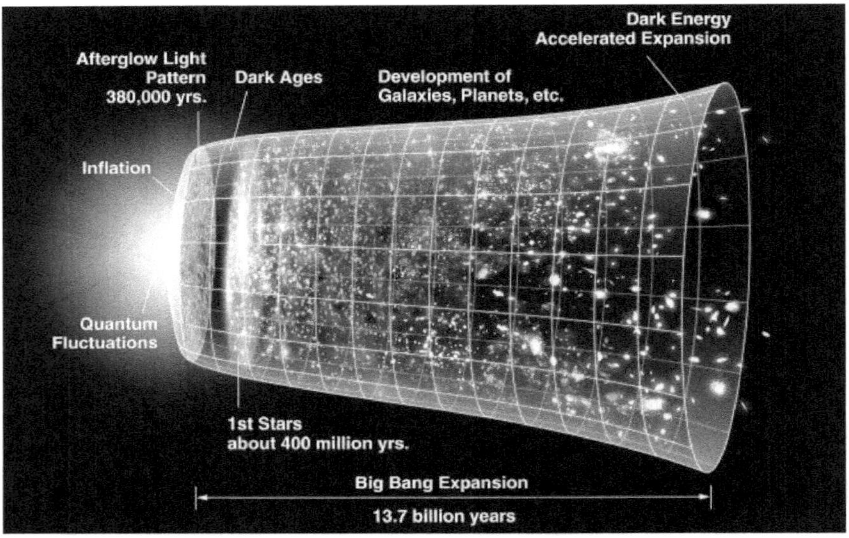

Possible Answers to the Thought Challenge

Understanding Hawking Radiation

Answer:

Hawking radiation presents a quantum twist to the traditional view of black holes as inescapable cosmic traps. Due to quantum fluctuations near the event horizon, particle-antiparticle pairs are generated. Here, one particle falls into the black hole, while its counterpart escapes, suggesting that black holes can emit radiation. This phenomenon challenges the classical understanding of black holes and underscores the complex relationship between quantum mechanics and gravity. It reveals that black holes are not just cosmic "sponges" absorbing everything but can also release radiation, altering our perception of their nature and impact on the cosmic landscape over vast timescales.

Quantum Insights Into the Big Bang

Answer:

Quantum spacetime challenges the classical cosmology view of the Big Bang as a singular, infinitely dense point. Instead, it proposes that spacetime consists of discrete units, suggesting the universe's inception was a series of smaller, quantized events rather than one abrupt explosion. This quantum perspective offers a vision of a less violent and more gradual cosmic beginning, where expansion and cooling occurred smoothly, sidestepping infinite densities and temperatures. This revised understanding could profoundly change our perception of cosmic events and timelines, akin to recognizing that each spark in a fireworks display originates independently, contributing to the overall spectacle in distinct, measurable ways.

Conclusion

We've reached the end of our exploration into the fascinating world of quantum physics. You've taken a step into a realm where particles can exist in two places at once, where seemingly distant objects remain mysteriously connected, and where the very fabric of reality bends to the rules of the quantum world.

Throughout this book, you've learned about the fundamental principles that shape quantum mechanics, from the strange behavior of particles in superposition to the mind-bending possibilities of entanglement and quantum tunneling. We've also explored how these principles aren't just abstract theories but are at the heart of the technology we use every day, from smartphones to medical imaging, and how they might drive the next revolution in computing and security.

You've seen how quantum mechanics doesn't just explain the smallest scales but pushes our understanding of the universe itself, bridging gaps between classical physics and the mysteries of relativity. Along the way, we've also ventured into some of the biggest questions that still remain, from paradoxes that challenge our understanding to the potential of a unified theory that may one day connect all the forces of nature.

I hope this book has made quantum physics more approachable and sparked your curiosity to continue learning. If you found it engaging or helpful, I'd love to hear your thoughts. A review would mean a lot and help others find their way into this fascinating subject.

As for what comes next, you might explore deeper into specific areas that caught your interest, stay updated on the latest quantum breakthroughs, or join communities of fellow enthusiasts to keep learning and sharing. Every new piece of understanding adds to the bigger picture, and there's always more to discover in the ever-evolving field of quantum physics.

Thank you for joining me on this exciting exploration. Who knows what new insights and breakthroughs await you as you continue to explore the wonders of science?

References

Alexander, S. (2023, October 28). *This bold new theory of "quantum weirdness" could rewrite the story of evolution.* BBC Science Focus Magazine. https://www.sciencefocus.com/comment/quantum-weirdness-force-of-life

Anastasiou, B. (2020). Quantum Field theory I. https://people.phys.ethz.ch/~babis/Teaching/QFT1/qft1.pdf

Ball, P. (2018). *Why the many-worlds interpretation has many problems.* Quanta Magazine. https://www.quantamagazine.org/why-the-many-worlds-interpretation-has-many-problems-20181018/

Crease, R. P. (2024, March 21). *Wigner's friend: The Quantum thought experiment that continues to confound.* Physics World. https://physicsworld.com/a/wigners-friend-the-quantum-thought-experiment-that-continues-to-confound/

Dai, X. (2020). The black hole paradoxes and possible solutions. *Journal of Physics: Conference Series, 1634*(1), 012088. https://doi.org/10.1088/1742-6596/1634/1/012088

Giustina, M., Versteegh, M. A. M., Wengerowsky, S., Handsteiner, J., Hochrainer, A., Phelan, K., Steinlechner, F., Kofler, J., Larsson, J.-Å., Abellán, C., Amaya, W., Pruneri, V., Mitchell, M. W., Beyer, J., Gerrits, T., Lita, A. E., Shalm, L. K., Nam, S. W., Scheidl, T., … Zeilinger, A. (2015). Significant-loophole-free test of Bell's theorem with entangled photons. *Physical Review Letters*, *115*(25). https://doi.org/10.1103/physrevlett.115.250401

Hilgevoord, J., & Uffink, J. (2016, July 12). *The Uncertainty Principle.* Stanford Encyclopedia of Philosophy. https://plato.stanford.edu/entries/qt-uncertainty/

Kemp, H. (2022). *Schrödinger equation.* Study. https://study.com/academy/lesson/schrodinger-wave-equation-overview-time-dependent-independent-equation.html

Kim, S., & Ham, B. S. (2023). Observations of the delayed-choice quantum eraser using coherent photons. *Scientific Reports*, *13*(1). https://doi.org/10.1038/s41598-023-36590-7

Oppenheim, J. (2023). A postquantum theory of classical gravity? *Physical Review X*, *13*(4). https://doi.org/10.1103/physrevx.13.041040

Sherrill, D. (2006). *The Ultraviolet Catastrophe.* The ultraviolet catastrophe. http://vergil.chemistry.gatech.edu/notes/quantrev/node3.html

Siegel, E. (2023, June 1). *Hawking radiation isn't just for black holes, study shows.* Big Think. https://bigthink.com/starts-with-a-bang/hawking-radiation-black-holes/

Styer, D. (1999). *A Brief History of Quantum Mechanics.* A brief history of Quantum Mechanics. https://www2.oberlin.edu/physics/dstyer/StrangeQM/history.html

Vyas, R. P., & Joshi, M. J. (2021). Loop Quantum Gravity: A promising candidate of theory of quantum gravity. https://www.researchgate.net/publication/353267848_LOOP_QUANTUM_GRAVITY_A_PROMISING_CANDIDATE_OF_THEORY_OF_QUANTUM_GRAVITY

Weinstein, G. (2023, October 6). Navigating the conjectural labyrinth of the Black Hole ... https://arxiv.org/pdf/2310.03607

Images

AartlistDesign. (2024, May 07). *Environmental technology concept* [Image]. Pixabay. https://pixabay.com/illustrations/environmental-technology-concept-8747662/

Aleserralvo. (2023, November 27). *Ai generated data technology* [Image]. Pixabay. https://pixabay.com/illustrations/ai-generated-data-technology-8412725/

Angela Compagnone. (2018, February 25). *Person holding smartphone taking video of people* [Image]. Unsplash. https://unsplash.com/photos/person-holding-smartphone-taking-video-of-people-4Iyg6cNU7sI

Cottonbro Studio (2018, January 17). *Person holding orange and white round ornament* [Image]. Pexels.

https://www.pexels.com/photo/person-holding-orange-and-white-round-ornament-4778677/

Geralt. (2018, December 10). *Schrodingers cat* [Image]. Pixabay. https://pixabay.com/illustrations/physics-schr%C3%B6dingers-cat-3864568/

Geralt. (2023, March 15). *Physics quantum physics particles* [Image]. Pixabay. https://pixabay.com/illustrations/physics-quantum-physics-particles-7854063/

Google DeepMind. (2024, June 04). *A black and white drawing of a ball and a circle* [Image]. Unsplash. https://unsplash.com/photos/a-black-and-white-drawing-of-a-ball-and-a-circle-qOOd_pM4jCk

Google DeepMind. (2024, June 04). *A black and white image of an American flag* [Image]. Unsplash. https://unsplash.com/photos/a-black-and-white-image-of-an-american-flag-GVGnKgEomlw

Janeb13. (2016, January 28). Albert Einstein and Niels Bohr casual [Image]. Pixabay. https://pixabay.com/photos/albert-einstein-and-niels-bohr-casual-1163429/

Kerplode. (2015, October 14). *White wall* [Image]. Pixabay. https://pixabay.com/photos/midway-game-midway-games-midway-983384/

Kier in Sight Archives. (2021, May 29). White and black abstract illustration [Image]. Unsplash. https://unsplash.com/photos/white-and-black-abstract-illustration-qXA4b_dZSbQ

Lindsey Garrett (2022, October 11). *Le passe muraille sculpture in Paris* [Image]. Pexels. https://www.pexels.com/photo/le-passe-muraille-sculpture-in-paris-13987246/

Nikita Khandelwal. (2018, January 28). *Shallow focus photography of person catching two gold colored coins* [Image]. Pexels.

https://www.pexels.com/photo/shallow-focus-photography-of-person-catching-two-gold-colored-coins-944650/

No name. (2018, April 07). *Atom proton electron neutron* [Image]. Pixabay. https://pixabay.com/illustrations/atom-proton-electron-neutron-3301567/

OpenClipart Vectors. (2013, July 11). *Scattering quantum mechanics* [Image]. Pixabay. https://pixabay.com/vectors/scattering-quantum-mechanics-145104/

Pietro Battistoni. (2018, January 17). *Accelerator* [Image]. Pexels. https://www.pexels.com/photo/806763/

Saylowe. (2017, March 14). *Orbitals bonding atom molecule* [Image]. Pixabay. https://pixabay.com/illustrations/orbitals-bonding-atom-molecule-2146393/

Steve Johnson. (2018, March 09). *White wall* [Image]. Pexels. https://www.pexels.com/photo/white-wall-927546/

WikiImages. (2011, December 14). *Galaxy big bang explosion space* [Image]. Pixabay. https://pixabay.com/photos/galaxy-big-bang-explosion-space-11188/

WikiImages. (2012, January 08). *Universe space expansion big bang* [Image]. Pixabay. https://pixabay.com/illustrations/universe-space-expansion-big-bang-11636/

www.ingramcontent.com/pod-product-compliance
Lightning Source LLC
LaVergne TN
LVHW022218200926
848481LV00006B/2014